The Earth's Human Carrying C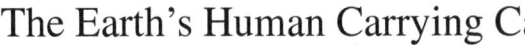

Frederic R. Siegel

The Earth's Human Carrying Capacity

Limitations Assessed, Solutions Proposed

 Springer

Frederic R. Siegel
George Washington University
Washington, DC, USA

ISBN 978-3-030-73478-7   ISBN 978-3-030-73476-3 (eBook)
https://doi.org/10.1007/978-3-030-73476-3

© The Editor(s) (if applicable) and The Author(s), under exclusive license to Springer Nature Switzerland AG 2021
This work is subject to copyright. All rights are solely and exclusively licensed by the Publisher, whether the whole or part of the material is concerned, specifically the rights of translation, reprinting, reuse of illustrations, recitation, broadcasting, reproduction on microfilms or in any other physical way, and transmission or information storage and retrieval, electronic adaptation, computer software, or by similar or dissimilar methodology now known or hereafter developed.
The use of general descriptive names, registered names, trademarks, service marks, etc. in this publication does not imply, even in the absence of a specific statement, that such names are exempt from the relevant protective laws and regulations and therefore free for general use.
The publisher, the authors, and the editors are safe to assume that the advice and information in this book are believed to be true and accurate at the date of publication. Neither the publisher nor the authors or the editors give a warranty, expressed or implied, with respect to the material contained herein or for any errors or omissions that may have been made. The publisher remains neutral with regard to jurisdictional claims in published maps and institutional affiliations.

This Springer imprint is published by the registered company Springer Nature Switzerland AG
The registered company address is: Gewerbestrasse 11, 6330 Cham, Switzerland

I dedicate this book to my grandchildren Naomi, Coby, and Noa Benveniste and Solomon and Beatrice Gold who will be living in a world with increasing limits on the earth's human carrying capacity to sustain a growing world population from 7.8 billion citizens in 2020 to 9.9 billion projected for 2050 because of global warming and its effects on water and food security and other basic needs plus impacts of extreme weather events. I dedicate the book as well to readers who will grasp the reality of such limits and their increasingly damaging socio-economic impacts during coming decades and work to have governments "follow the science" and put in place with urgency adaptations and projects to slow and rein in global warming.

Preface

There are limits to how large a human population the Earth has the capacity to support and sustain with a good quality of life. The limits are set mainly by population growth, availability of water, agriculture production, depletion of non-renewable energy and other natural resources, industrial output, and pollution. The limits may be extended by human capabilities to undertake adaptations to damp the negative effects of changing environmental conditions. These adaptations have expanded the useful life of resources and created others that are helping to meet basic needs for all of our Earth's population and that before COVID-19 were reducing global poverty. This book examines the conditions of the Earth's 7.8 billion citizens in 2020 and those of a population growth projected to be 9.9 billion people by 2050, and possibly higher yet later in the century. These are availability and access to clean water, nutritious food, sanitation status, and natural resources, all evaluated as Earth ecosystems suffer the effects of progressive global warming and changes in climates it causes. This book considers how humanity is responding to population growth and meteorological changes by adopting policies and methodologies that sustain and improve human existence, thus providing the essentials of life and added benefits by bettering socio-economic situations that ultimately serve all societal classes. There are many numerical values in the text and accompanying tables for you, the reader, to digest with respect to the topic(s) being discussed. Some are in place to set an order of magnitude of measurements, others to show the tendency of change with time, and still others that are computer generated for predictive purposes into the twenty-first century that will change with revised input values. Keep this in mind. Lastly, there is repetition among chapters but repetition with added discussion. Good reading.

Washington, DC, USA   Frederic R. Siegel

Contents

1. **Earth's Human Carrying Capacity: The Basics** 1
   References 4
2. **Water: A Limit on Our Earth's Carrying Capacity** 7
   2.1 The Water Cycle Is Changing 7
   2.2 International Water Stress 8
   2.3 Desalination 10
      2.3.1 Desalination Problem: Disposal of Brine 11
      2.3.2 Ecological Problem 13
   2.4 Reuse of Wastewater 13
   2.5 Import/Transference of Potable Water 14
      2.5.1 Pipelines 14
      2.5.2 Pipelines for Potable Water Transport 15
      2.5.3 Aqueducts 16
      2.5.4 Other Possibilities to Supply Potable Water to Areas Suffering Severe Water Shortages 17
   2.6 Some Numbers on Earth's Fresh Water Carrying Capacity for Human Needs 17
   Afterword 19
   References 19
3. **Food Security/Insecurity, Food Systems** 21
   3.1 Introduction 21
      3.1.1 Food Insecurity 21
   3.2 Soil: A Foundation of Food Security 22
      3.2.1 Minimizing Soil Erosion 24
      3.2.2 Maintaining Soil Nutrient Contents 25
   3.3 Population Challenge to Food Security 26
   3.4 Food Security Dependency on Food Systems 27
   3.5 Food Production: The Basis for Food Security 28
      3.5.1 Increasing Food Production 29
      3.5.2 Speed Breeding by Extended Photoperiods Exposure Adjunct to Hybridization 29

ix

		3.5.3	Mutation by Space Exposure to Cosmic Radiation	30
		3.5.4	Expansion of Farmland	30
		3.5.5	Population Growth and Farmland Expansion	31
		3.5.6	Improve Food Security: Reduce Food Loss, Food Waste	32
	3.6	Adding Farmland from Within for Food Production		34
		3.6.1	Reclaim Cropland from Fuel Feed Stock	34
		3.6.2	Planting GMO Crops Can Free Land for Additional Food Production	35
	Afterword			35
	References			36
4	**Impact of Global Warming/Climate Change on Food Security 2020**			**39**
	4.1	Introduction		39
	4.2	The Warming Status		39
	4.3	Extreme Weather Disasters Impact on Food Production		40
		4.3.1	Examples of Extreme Weather Disasters and Their Effects on Food Supplies in Recent Decades	41
	4.4	Observed and Measured Ongoing Effects of Global Warming on Ecosystems and Food Production		43
	4.5	Counter the Effects of Weeds, Insects/Pests, Disease		44
	4.6	Warmer Temperature Effects on Crop Physiological Growth		44
	4.7	Survival Plan: Reduce CO_2 Emissions		45
		4.7.1	Removal of CO_2 from the Atmosphere?	46
	Afterword			47
	References			47
5	**Sanitation, Waste Generation/Capture/Disposal Status 2020**			**49**
	5.1	Introduction		49
	5.2	Open Defecation: A Problem and a Solution		49
		5.2.1	Reducing Open Defecation with Education	50
	5.3	Food Animal Wastes		51
		5.3.1	Magnitude of the Global Food Animal Waste Problem	52
		5.3.2	Using Slaughter House Wastes	53
	5.4	Pollution (Waste) Endangers Earth's Human Carrying Capacity		54
		5.4.1	Can We Sanitize or Must We Depopulate Areas Evacuated During/After a Major Radiation Release?	55
		5.4.2	CO_2, a Pollutant?	56
	Afterword			57
	References			57
6	**Access to Natural Resources Not Water or Food 2020**			**59**
	6.1	Introduction		59
	6.2	Human Resources		59
	6.3	Energy and Electricity		61

	6.4	Metals	64
	6.5	Industrial Rocks and Minerals	66
	6.6	Wood	66
	6.7	Population Growth, Poverty, and Consumption	68
	Afterword		69
	References		69

7 Global Warming and Water 2050: More People, Yes; Less Ice, Yes; More Water, Yes; More Fresh Water, Probably; More Accessible Fresh Water? 71

	7.1	Temperature Conditions	71
	7.2	Fresh Water from Ice: Water Towers	72
		7.2.1 Some Consequences of Declining Meltwater Flow from the Alps	73
		7.2.2 Some Consequences of Declining Meltwater Flow from the Andes	75
		7.2.3 Glacial Ice Thinning	75
	7.3	Changes in the Water Supply/Security Exclusive of Mountain Glaciers	76
	7.4	Future Regional Water Availability (Increase, Stable, Decrease) with Progressive Global Warming Through the Twenty-First Century	77
	7.5	Water at Risk/Water Availability/Water Food/Water Hazards as Twenty-First Century Progresses	79
		7.5.1 Threats to Coastal Cities and Aquifers	80
		7.5.2 Rainfall Deficit Impact on Hydropower	81
		7.5.3 The Ethiopian Dam Threat to Egypt's Water Lifeline: 2021 Conflict Resolution Necessary	82
	7.6	A Solution to Bring Water to Water-Starved Populations	82
	Afterword		84
	References		84

8 Food 2050: More Mouths to Feed—Food Availability and Access 87

	8.1	Introduction	87
	8.2	The Challenges	88
	8.3	Effects of Continued Climate Change on Food Production	88
		8.3.1 Projections for Future Food Production	89
		8.3.2 Fisheries	93
		8.3.3 Warming Temperature Effects on Food Fish	94
	8.4	Other Warming Threats to the Food Fish Supply/Security	95
	8.5	Hypothetical Solution to Feed Growing Earth Populations	96
		8.5.1 Diet Change = More Land for Food Production	96
		8.5.2 Healthy Diet for Ten Billion People in 2050?	97
		8.5.3 In Theory, in Practice	98
		8.5.4 Diet Boost from Kitchen Prepared Insects?	99

	8.6	Some Foreseen Effects of a Rising Global Temperature	99
		8.6.1 Reject Coal or Not in Industrial Plants or as Export?	100
		8.6.2 Forecast of Regional Temperature Effects	101
	8.7	To Slow, to Stop, to Reverse Global Temperature Rise	103
		8.7.1 Direct Extraction of CO_2 from Air	104
	Afterword		106
	References		106
9	**Sanitation 2050**		109
	9.1	Introduction	109
	9.2	Open Defecation: A Threat to Public Health—A Solution	109
	9.3	Waste Disposal Management	111
		9.3.1 Landfills	111
		9.3.2 Incineration	112
	9.4	Nuclear Waste: A Hazard in Search of Secure Disposal	112
	Afterword		114
	References		115
10	**Natural Resources Beyond Water and Food 2020–2050**		117
	10.1	Energy Resources	117
		10.1.1 Fossil Fuels	119
		10.1.2 Solar and Wind Electricity Sources	121
		10.1.3 Nuclear Power Plants	121
		10.1.4 Biomass	122
		10.1.5 Battery EVs, HEVs, PHEVs	122
	10.2	Critical/Strategic Metals, Industrial Minerals and Rocks	123
	10.3	Loss of Forestland or Not? An Important CO_2 Sink, a Source of Timber, Food, and Medicinal Plants	125
		10.3.1 Negative Use Impacts, Positive Use Benefits	125
		10.3.2 Forest Loss	127
		10.3.3 A Declining Rate of Forest Loss	127
		10.3.4 Forest Gain?	128
		10.3.5 Illegal Logging	129
	10.4	Forestry Management	130
		10.4.1 Guidelines for Sustainable Logging	130
		10.4.2 Logging Systems	131
	10.5	Preserving the Future: Human Resources	131
	Afterword		133
	References		133
11	**Economic Realities in 2020 Populations: What Do They Portend for 2050? 2100?**		135
	11.1	Introduction	135
	11.2	The COVID-19 Effect	136
	11.3	Unemployment, Underemployment: A Global Problem	137
		11.3.1 Poverty/Extreme Poverty	138

		11.3.2 Extreme Poverty by World Bank Income Levels	139
		11.3.3 Status of Global Extreme Poverty	140
	11.4	Global Economic Power Shift	141
	References		145

Epilogue 147

Index 149

Chapter 1
Earth's Human Carrying Capacity: The Basics

The Earth's carrying capacity is best defined as the maximum population size of a species that an area (environment, ecosystem) can support without reducing its ability to support the same species in the future (indefinitely). The Earth has limits to the life it can support, it's true carrying capacity for the intertwined relations among humans, animals, and plants in the varying ecosystems they inhabit. The basic limitations are: water and food for all life forms, soil, water, and sunlight to grow food for all life forms, and viability of ecosystems with natural resources that help sustain human, animal, and vegetation populations.

Many researchers have developed analyses that have allowed them to give their estimates of the earth's human carrying capacity. A 1995 report considered historical and modern studies on the subject and found estimates of the Earth's human carrying capacities ranging from populations of <1 million to >1000 billion people [1, 2]. Evaluation of the mean highs and mean lows in 65 reports yielded predicted low and high carrying capacities for **2050 of 7.5–12.5 billion people**.

These population numbers were based on global fertility rates that were in decline and a relatively unchanged rate of deaths. Global fertility rates have decreased gradually from 4.74 in 1970 to 2.35 in 2020 but the number of global births annually has been on the rise whereas global death rates have had a much lesser rate of increase with the result that from about 1975, the global populations have been growing by 75 to >80 million people yearly (Table 1.1). The lack of great change in the death rates with respect to the birth rates that previously reduced annual global population increases is attributed to better health care and improving life styles vis a vis, for example, education, increased exercise, better dietary choices, reduced smoking, and desirable social interactions.

The result is that **in mid 2020, the global populations was ~7.8 billion people and increasing**. Demographically, the city populations totaled **~4.2 billion** persons with **~3.6 billion** in rural areas. The present projection for Earth's **2050** human population is **~9.9 billion people** with **~6.9 billion** living in cities and **~3 billion** in rural settings [4].

Table 1.1 Changes in global births and deaths since 1950 in 5 year intervals together with their ratios and numerical increases in the world population that year. Modified from: ourworlddata.org/grapher/births-and-deaths [3]

Year	Births/deaths in millions	Ratio	Real growth in millions
1950	97.40/51.27	1.90	46.13
1955	99.44/50.36	1.97	49.08
1960	107.54/51.25	2.10	56.29
1965	116.77/49.60	2.35	67.17
1970	121.50/46.76	2.60	74.74
1975	121.56/46.04	2.64	75.52
1980	125.15/46.25	2.70	78.90
1985	135.97/47.63	2.85	88.34
1990	138.23/49.43	2.80	88.80
1995	132.06/51.23	2.58	80.83
2000	130.96/52.60	2.48	78.36
2005	134.72/53.49	2.52	87.23
2010	138.28/54.17	2.55	84.11
2015	140.87/56.66	2.49	84.21
2019	140.79/59.43	2.37	81.36
2020[a]	140.66/60.20	2.34	80.46

[a]Projected

Table 1.2 World and regional populations distribution in mid-2020 and projected to mid-2035, and mid-2050, in billions [4]

	Mid-2020	Mid-2035	Mid-2050	% Change 2020–2050
World	7,773	8,937	9,876	27
Africa	1,338	1,897	2,560	91.3
Asia	4,626	5,112	5,331	15.2
Europe	0,747	0,744	0,729	−2.4
Latin America and Caribbean	0,651	0,724	0,759	16.6
Northern America	0,368	0,406	0,435	10.2
Oceania	0,043	0,053	0,063	46.5

The mid-2020 populations for the world and its principal regions, and projections to mid-2035 and mid-2050 are presented in Table 1.2 together with the percent increases from 2019 to 2050. It shows an astounding almost doubling of the African population in 31 years and a negative population growth for Europe. It is difficult to envision an Earth's carrying capacity that is able to sustain the 2050 African population when many of the continent's 2020 populations, especially in Sub Saharan Africa, suffer from water and food deficits and insufficient healthcare services.

The question that has to be considered is the relation of carrying capacity to the global population socio-economic structure…the reality of sustainability. Do we determine carrying capacity differently for "upper class, middle class, lower class populations?" Is there a defined carrying capacity for "economically advantaged segments of society, for economically comfortable populations, for economically disadvantaged and poverty groups?"

If asked the question of whether the Earth's human carrying capacity **is sustaining** the 2020 world population, the answer has to be **no**. If asked the question of whether or not the Earth's human carrying capacity **can sustain** the 2020 global population, the answer has to be **'perhaps'** to **'yes'** depending on the economics of doing so and the peace/no peace/conflict/war situation in many regions.

Calculations/estimations of the Earth's human carrying capacity has many parameters to evaluate in seeking a viable answer for 2050 or further in the future. Primary among these are projected population growth and demographics between urban and rural populations as cited in an earlier paragraph. These population growth figures set the societal needs for basic sustainability with respect to water and food security plus sanitation, and the Earth's waste disposal absorption capacity. Added to these are accessibility to natural resources in addition to water and food, plus an equilibrated interaction with the living environment so that what is harvested and used does not exceed what is replaced. Biodiversity should be protected with the preservation, for example, of inland and coastal agricultural terrain, and ocean/inland fisheries/aquaculture. These, plus medical care, technological advances, and corresponding economic development can keep people alive and societies functioning. All such factors are at risk from existing conflicts/wars and epidemic health threats in some global regions or pandemics that present threats worldwide as in 2020/2021, both of which can unfortunately be heightened by imprudent political decisions by one or more nations or by poorly informed or politically motivated heads of state. All of the above factors, plus others that will be discussed in later chapters, have to be assessed in 'now' and future (2050, 2100) planning for the existing and increasing effects of global warming and resulting climate changes that impact people, the environments they inhabit, and ecosystems that support them.

With respect to the above discussion, we have to ask, where does the world stand with respect to its human carrying capacity and the 2020 global population? Is it likely that the carrying capacity has already passed its tipping point? We know that of the ~7.8 billion people on Earth, safe water is a problem for one billion people (~13%) or if sanitation is considered, for two billion people (~26%) and food is a problem for more than two billion (again ~26%). The water and food deficits exist now. Can these deficits be erased and life support needs be extended to service a projected population of ~9.9 billion persons and allow all to have a reasonable quality of life? If so, how? If not, what does the future portend?…famine migrants, drought migrants, economic migrants, from where to where, wars over natural resources?

More than 821 million people suffered from hunger, food insecurity and malnutrition worldwide in 2018, a figure that was in decline years previous to 2015 but since then has been on the increase. Malnutrition, an insufficient intake of protein energy and micronutrients (e.g., vitamin A, iron, iodine) remains widespread, especially in Sub-Saharan Africa with ~20% of the populations affected and in Asia where >12% of the population suffer malnutrition. About 7% of the populations in South and Central America are affected by malnutrition. Consider with these the number of people experiencing famine and food insecurity and there is an estimated total of >2 billion people that suffer from lack of availability, access to, or reliability of food supplies. Of the 821 million suffering from malnutrition, there are 149 million children under 5 years old (~20% of the total worldwide) for whom malnutrition has resulted in stunted growth and impaired cognitive abilities. India alone accounts for >48 million of the 149 million affected children [5, 6]. This 2020 problem can be solved over time with directed planning. However, can it be solved for a 2050 estimated ~2.3 billion more people globally, mainly in Africa and Asia, a little more than a generation away? This, and factors noted earlier, and others that are fundamental in determining the Earth's human carrying capacity will be discussed in the following chapters with a ~2020 baseline and assessments of future possibilities that exist for increasing the Earth's carrying capacity to cope with future population growth.

There is a breadth of meaning to the Earth's human carrying capacity that is determined mainly by persons' socio-economic status. This is true whether in democratic, socialist/communist, or autocratic/despotic societies and is muted to some degree by a government's welfare or basic support for economically disadvantaged citizens in a city/country. Economically advantaged groups in a society will be heavy consumers of water, food, and natural, municipal, and utilities resources (electricity, natural gas) than will be groups not so advantaged. Thus, an addition to the definition of the Earth's human carrying capacity is necessary and would mean supplying **to all human populations**, at the least, clean water for drinking, cooking, and personal hygiene, food that nourishes them with the protein and nutrient needs to keep people healthy, sanitation systems including waste collection and its safe disposal that also abets human health, and access to good quality medical attention. Added carrying capacity elements include access to basic natural resources that sustains other human needs, electricity, and educational opportunities. Together, these elements and other considered in following chapters will contribute to poverty reduction and self-sufficiency. Because several of these are lacking in 2020 for many populations worldwide, especially in countries in Africa, Asia, and somewhat in Latin America, we can ask what is the likelihood that existing negative conditions will change towards the positive if a projected ~2.3 billion more people inhabit the Earth in 2050. This is a special challenge in those countries/regions where, for all citizens in 2020, the basic human carrying capacity requirements cited above are not being met!

References

1. Cohen, J.E., 1995. Population growth and the earth's human carrying capacity. Science, 269: 341-346. https://doi.org/10.1126/science.7618100
2. Cohen, J.E., 1996. How many people can the Earth support. Norton & Company, NY, 532 p.

References

3. Online. worldmeters.info/world-population, 2020. Past, present, and future.
4. Population Reference Bureau, 2020. World Population Data Sheet. Washington, D.C.
5. Food and Agriculture Organization of the United Nations, and others, 2019. The State of Food Security and Nutrition in the World. Rome, 212 p. (121 p text, 22 p. notes, 69 p., annexes).
6. Myers, S.S., Smith, M.R. and 6 co-authors, 2017. Climate change and global food systems: potential impacts on food security and undernutrition. Annual Reviews of Public Health, 38: 259-277.

Chapter 2
Water: A Limit on Our Earth's Carrying Capacity

Introduction Water is the essence of life. Water hydrates humans, sustains agriculture and animal husbandry that feed the Earth's growing populations, is the medium that supports personal hygiene and for sanitation systems that are the keystone to public health, and is basic to industrial development. Although there is enough water on our planet to serve the needs of all populations, it is irregularly distributed geographically. The result is water stress that increases with growing populations and intensifies per capita for billions of people in water deficit regions. In some instances, water stress can be eased by desalination that can disrupt ecosystems and in other cases with aqueduct systems, but not for all. These topics and others related to water transference to sustain populations in areas under water stress are considered in the following paragraphs.

2.1 The Water Cycle Is Changing

The Earth's total surface water/groundwater volume is relatively stable but has been partitioning differently, albeit slowly, between water in the oceans, water locked up as ice in continental and mountain glaciers, ice sheets and permafrost, as well as water flowing in rivers and streams, water in swamps and stored in lakes, and as moisture in soils, and humidity in the atmosphere. These changes of environments the H_2O partitions from and to is the result of global warming. Table 2.1 shows the percentage of the Earth's surface and groundwater for each.

The values given in Table 2.1 are changing, albeit slowly, but changing perceptively as a result of global warming. There is less ice locked up in icecaps, mountain glaciers, and ice sheets as melting is taking place worldwide. During 2019, a much warmer than usual summer temperature in Greenland of 22 °C (71.6 °F) caused glacier melt of 586 billion tons of ice (~596 km^3) into the North Atlantic that resulted in a global sea level rise of 1.5 mm. The fact is that global warming evaporates more moisture into the atmosphere where clouds carry the water to precipitate, some back into the ocean and some inland where more water (than normal in past decades) from storms runs off and fills rivers and streams and lakes they may run into. The precipitation may also seep into soils and ofttimes through them and underlying

© The Author(s), under exclusive license to Springer Nature Switzerland AG 2021
F. R. Siegel, *The Earth's Human Carrying Capacity*,
https://doi.org/10.1007/978-3-030-73476-3_2

Table 2.1 Locations and percentage volumes of the total Earth surface and groundwater water inventory [1, modified]

Location	% Total water
Oceans and seas	96.5%
Sources for desalinization	
Fresh water	3.5%
Ice caps, mountain glaciers and ice sheets	1.74%
Ground ice and permafrost	0.002%
In surface waters	0.014%
Fresh water lakes	0.007%
Rivers and streams	0.0002%
Swamps	0.0008%
Subsurface fresh water	
Fresh water aquifers	0.76%
Shallow	0.34% <750 m
Deep	0.42% >750 m
Subsurface brackish water	0.93%
potential for desalinization and specific crop irrigation	
function of salinity: barley, asparagus, spinach, date palms	
Soil moisture	0.001%
Saline lakes	0.006%
Atmosphere	0.001%
Biological systems	0.0001%

rocks to recharge unconfined aquifers. Of all the H_2O on Earth, less than 1% is readily available to support humans and other life forms. The volume of the earth's liquid freshwater in rivers and lakes that people use daily is 93,113 km^3, whereas the volume in groundwater is 10,530,000 km^3, but much of which is too deep in the ground to be used by people [2].

2.2 International Water Stress

A 2019 paper reports that close to a quarter (23.3%) of the world's population lives in 17 countries with 'extremely high water stress' (20% attributed to India and Pakistan). Another 27 countries are subject to 'high water stress' while 23 countries have 'medium high water stress' and 95 countries, mainly in Europe, Africa, Asia, and Central and South America have 'low medium to low water stress' [3]. As understood from this classification, stress is a relative term, relative to what people need to fill the basic requirements for a healthy life. For daily requirements the World Health Organization (WHO) sets as a daily minimum, 3 l for drinking, 4 l for cooking, 20 l for bathing, and 40 l for sanitation, plus another 8 l for washing cooking vessels, for a total of 75 l (27.4 m³ annually) to fully protect against disease. For example, of the 1.1 billion people without access to any type of improved (safe/clean) drinking water, 1.6 million die annually from diarrheal diseases including cholera with 90% of these being children under 5 years old. A United Nations (UN) study reports that as many as 4.2 billion people of the ~9.9 billion projected 2050 population will be living in countries that can not meet the 75 l daily minimum [4]. The world average water use far exceeds this at 173 l per person per day with the United States daily use more than doubling this at 380 l. Table 2.2 shows the renewable water resources for selected countries, their per capita allotments, and the degree of water stress the their citizens experienced in 2018 and what can be projected for their 2050 populations.

The per capita designation is a volume that is partitioned into percentages used for the agricultural sector, the industrial sector, and for municipal/domestic uses.

Table 2.2 Examples of countries' renewable water resources and their per capita volume changes from 2018 projected to 2050

Country	Total renewable water resources (km³)	Per capita 2018 (m³)	Per capita 2050 (m³)
Grave			
Saudi Arabia	2.4	73	53
Algeria	11.67	278	182
Egypt	57.5	592	345
Kenya	30.7	602	318
Morocco	29.1	827	667
Somalia	14.7	967	409
Marginal			
Nigeria	286.2	1461	697
Ethiopia	122.1	1136	640
Uganda	60.1	1363	629
India	1911	1394	1137
Iran	137	1679	1471
Tanzania	96.27	1629	697
Water sufficient			
Russia	4525	30,719	32,814
United States	3069	9356	7877
Canada	2902	77,180	61,876
China	2840	2038	2113
Indonesia	2019	7613	6315

The data indicate a population's water stress level. Water stress is grave with <1000 m³ per capita annually, marginal with >1000–<2000 m³ per capita annually, and water sufficient with >2000 m³ per capita annually [2, 5]

Note: 1 km³ = 1 billion m³

Table 2.3 Examples of percentage of water use by sector for countries listed in Table 2.2 illustrating the country differences between agricultural dominated economies, industrial economies, and municipal requirements [6]

Country	Agriculture	Industrial	Municipal
Saudi Arabia	82.23	4.28	13.49
Algeria	64.14	1.88	33.97
Egypt	79.16	6.97	13.87
Kenya	80.21	7.51	12.28
Morocco	87.79	2.03	10.19
Somalia	99.48	0.06	0.46
Nigeria	44.19	15.76	40.1
Ethiopia	89.05	0.65	10.31
Uganda	40.66	7.85	51.49
India	90.41	2.23	7.36
Iran	92.18	1.18	6.64
Tanzania	89.35	0.48	10.17
Russia	26.24	49.53	24.23
United States	39.66	47.3	13.14
Canada	7.38	78.56	14.05
China	64.4	22.32	13.28
Indonesia	85.88	4.10	10.69
World Usage	70.00	22.00	8.00
Germany	2.36	19.95	77.69
Japan	66.83	14.25	18.92
Viet Nam	94.78	3.75	1.47

Table 2.3 illustrates the percentages found for the countries listed in Table 2.2, most with growing populations. It is clear that for 9 of the 12 countries listed in Table 2.3 with grave to marginal per capita water availability in 2018, that agricultural projects for domestic food needs uses more than about 80% of their total renewable water resources whereas industrial water use accounts for less than 8% for 11 of the 12 nations (with 3 <1%). In these groups, Nigeria presents the more balanced use of water resources. For the water sufficient group, Russia (with a shrinking population) and the United States are more balanced in their water used for the three sectors whereas Canada uses almost 80% for industrial projects, with China (very slowly contracting population) and Indonesia devoting almost 65% and 86%, respectively, to food production.

2.3 Desalination

Desalination adds to the Earth's inventory of clean water. There were some 16,000 desalination plants operational during 2017 that produced 95.37 million m³ daily (0.09533 km³) almost half of which serves the Middle East and North Africa

2.3 Desalination

Table 2.4 Global desalination output of potable water during 2018 by region in millions of m³ per day [8]

Region	Capacity
Middle East and North Africa	45.32
East Asia and Pacific	17.52
North America	11.34
Western Europe	8.75
Latin American and Caribbean	5.46
Southern Asia	2.94
Eastern Europe and Central Asia	2.26
Sub Saharan Africa	1.78

(Table 2.4). In these regions, rainfall is little, evaporation rates are high, and aquifers are in decline because of lack of recharge. Without the desalination input of clean water, some nations would not be able to support their populations. For example, 30 desalination plants in Saudi Arabia (35 million people) supply >70% of the population's potable water needs, 7 plants in the UAE (9.8 million people) and 3 in Qatar (2.7 million people) supply their populations with >90% of the potable water needs, and 6 desalination plants in Kuwait (4.7 million people) provide >95% of that country's potable water supply. Clearly any conflict or expanded conflict in this region can and likely would threaten the potable water supply.

Saudi Arabia has invested US$billions to build 1000s of km of pipeline from desalination plants and reservoir dams to users. For example, there is a 467 km pipeline from Ras al Khair to Riyadh that is built for redundancy so that there are parallel pipelines for a total of 934 km at US$1.2 billion. This is one of about 27 desalination facilities each with pipelines to user populations. The economy, availability of cheap oil to produce the energy for desalination plants, and need of water for survival as its population grows, drives Saudi Arabia to invest in increasing its desalination capacity.

Of 17 countries that suffer from **extremely high water stress**, several have access to an ocean and could invest in increasing whatever desalination capacity they may require. These include Iran, Libya, Eritrea, India, Pakistan, and Turkmenistan. Similarly, several countries classified as having **high water stress** and with access to an ocean or sea can invest in desalination plants to ease the stress [7].

2.3.1 *Desalination Problem: Disposal of Brine*

In advanced desalination plants, 2 m³ of brine are generated to produce 1 m³ of potable water. As a result 141.5 million m³ of brine waste pollutants were generated in 2018 to produce 95.37 million m³ of potable water [9]. About 45% of the brine is from desalination plants in Saudi Arabia, Kuwait, the UAE, and Qatar and account for 55% of the global production of brine waste water [10]. The brine has 1.5× to close to 2× the normal sea water salinity (3.5% salt content by weight). This means that what has to be used or disposed of annually ranges from 7.64 to 10.18 million metric tonnes of toxic containing reaction byproducts, mainly NaCl but also chemicals used to clean bacterial growth and anti-scaling chemicals plus elements naturally in seawater and heavy metals from corrosion of machinery and pipes.

Brine from desalination plants using seawater as feed stock is being disposed of in different ways that are costly and damaging to the receiving environments/ecosystems. The brine may be discharged through outfall pipeline directly into the ocean. It is denser than normal seawater and if discharged as a plume will sink to the seafloor harming marine life. In some cases the brine is diluted to reduce its hyper salinity previous to discharge. In other cases, the brine has been discharged into fast moving currents with the expectation that it will be rapidly diluted to tolerable levels for marine life. In still other cases, the brine is moved to land to dry for possible use. Neither of these disposal methods is good for the environment and can harm the yield of animal and plant natural resources useful to societal food security [11]. How this mass is managed today and in the future with respect to disposal or use options has to be evaluated with respect to environmental/ecosystem health and economic ramifications.

The answer to the problems that can be caused by existing environmentally damaging brine disposal methods is to reduce the mass that is discharged by extracting useful products from the brine such as sodium chloride to sell as a de-icing agent and, as proposed by some, salable chemical elements as may be present in the brine and useful to industries such as bromine, lithium, potassium, boron, copper and others, including uranium. This is done with brine from the Dead Sea by the Israel Dead Sea Works with the extraction of potash and sale of potash products as well as magnesium chloride, industrial salts, deicers, bath salts, table salt, and raw materials for the cosmetic industry. The Jordan Bromine Company is a leading producer of bromine and bromine derivatives for the personal health care industry and other sectors. Although the extraction potential exists for desalination brines, the extraction of these 'use' chemical elements can be costly and most can not compete with existing commodity prices. This may change later in the century.

An innovative approach to dealing with the brine disposal problems has been tested in a chemical engineering lab at Qatar University by Professor Farid Benyahia with the aim of reducing excess CO_2 in the atmosphere and the discharge of excess salt (brine) from desalination plants into the Arabian Gulf [12]. The proposed approach is designed to yield useful products and offset some of the costs of capturing and storing CO_2. The main product manufactured is sodium bicarbonate ($NaHCO_3$, baking soda, a compound with many household uses). To begin with, in the presence of ammonia ($NH_{3(g)}$), pure CO_2 (no other gas impurities) reacts with the waste brine creating solid baking soda and ammonium chloride (NH_4Cl) solution. Next, the sodium chloride (NaCl) solution reacts with calcium oxide (CaO) to yield calcium chloride ($CaCl_2$) solution and ammonia gas for reuse in the first reaction. If applied in the Arabian Gulf, where we noted that about half of the global brine waste from desalination plants is generated, the pure CO_2 would come from a Qatari natural gas processing plant close to brine disposal stations. In theory, the process could reduce brine disposal in this region almost completely. Locations of desalination plants with a similar infrastructure (i.e., pure CO_2 generated near brine disposal stations) would be favorable cost-wise. Otherwise, implementation of the process would likely be too costly because of the price of the pure gas. This could change in the future as carbon capture systems are imposed on industries that emit CO_2 or by the direct capture of this gas from the air (see Sects. 4.7 and 8.7) and an economically viable infrastructure is developed to transfer the captured gas to desalination plants for the process described above.

2.3.2 Ecological Problem

It is important to note that other ecological problems are caused by desalination plants and these are being addressed in different ways with different degrees of success. One problem is that sea water intake pipes suck in and kill marine life such as fish, crabs, roe and plankton [11]. This problem can be mitigated by intaking sea water from deeper water rather than surface water, by lowering the velocity at which the water is taken in, and by using a finer opening screen for the intake pipe. Such changes, if viable, care reduce the impact on marine life and food web. Another problem is from thermal desalination plants that discharge heated water into the sea that both alters the ecosystem equilibrium and reduces oxygen (O_2) in the water, both which put life forms in the receiving waters at risk. Viable solutions are to cool the water or mixing it with normal temperature sea water before releasing it into the sea. The latter solution applies to brine that can be mixed with treated waste waters and discharged farther into the sea preferably into faster moving currents that dilute and bring them into environmental equilibrium.

2.4 Reuse of Wastewater

The reuse of domestic and industrial waste water treated to acceptable drinking standards for municipalities and to acceptable industrial uses is becoming an important contributor to the global water inventory. The United Nations estimates that high income countries treat 70% of municipal and industrial wastewater. Upper middle income and lower middle income countries treat 38% and 28% of this wastewater, respectively. Lower income countries treat only 8% of their municipal and industrial wastewater. It is environmentally sound to do so since as much as 80% of all wastewater and any contained toxins is discharged into rivers, lakes, and oceans [13]. Thus, a reduction of untreated wastewater discharge reduces ecosystem pollution hazards. This is most readily done where infrastructure (sewer systems) is in place to collect and move wastewater to existing or upgraded treatment plants and releasing safe water into a distribution network.

In such a scenario, treated wastewater reuse has supplemented a city's water supply to deal with a drought caused shortfall. This was the case in Melbourne, Australia during the 2000s Millennium drought when treatment plants processed 'gray' waters to supplement the water needs of its 1.1 million citizens. Reuse of treated wastewater has supplemented water supplies for agriculture in European Union nations and in Central and South America and Mexico. Although desalination is the principal source of potable water in the Middle East, interest in reuse of wastewater to improve water security has the UAE and Egypt upgrading sewage treatment plants for that purpose [14]. Many municipalities without the infrastructure and treatment facilities have begun investing in their construction or are in a planning stage to do so. Industrial demand for water of a certain volume and purity and the prioritized needs of municipalities with growing populations are driving industries to adapt for treating and recycling cleansed wastewater internally. This extends water supplies and hence a local/regional water carrying capacity.

2.5 Import/Transference of Potable Water

2.5.1 Pipelines

There are about 3.5 million km (2.2 million mi) of pipeline worldwide, mainly to move oil and natural gas from port sources or production venues to refineries and then within countries for domestic and industrial use. Of 123 countries listing pipelines, 4 have >100,000 km (62,111 mi) of pipelines (Table 2.5). Another 8 countries have >30,000 km (18,633 mi) of pipelines, and 11 have more than 20,000 km (12,423 mi) of pipelines [15]. The length of such pipelines in a country is a function of the size of a country, its economy, and the number of households and other domestic users and industrial users that are being served.

A country's pipeline inventory is a function of its contracts with suppliers of these energy resources moving systems that cover their installation and maintenance. Investment is made for profit from the sale of high value/high volume use of company products upon pipeline installation completion and in the future. Given that a barrel of crude oil has a 11 January, 2021 value between about $52 and $55 (depending on location and quality) whereas an equivalent volume of potable water has a value of ~$1 in Washington, DC, we have the profit motive for constructing pipelines to transport oil (and natural gas) large distances and not water. Clearly

Table 2.5 Countries with more than 100,000 km of pipeline to carry oil and natural gas. They represent ~85% of global oil and natural gas pipelines [16]

Country	km of oil/natural gas pipeline
United States	2,225,032
Russia	259,913
China	126,000
Canada	100,000

2.5 Import/Transference of Potable Water 15

there is no price point that will dictate when there can be construction of pipelines to export water from one country or one region to another, but rather the drive for survival to service existing populations and growing populations in the future will demand government investment. Otherwise, the failure to invest could result in mass deaths from lack of potable water and water to support food production. Failure to come to grips with this water problem reflects economic conditions and politics in less developed and developing countries that are under water stress and/or with relations between nations or regions where excess water may be available for export.

2.5.2 Pipelines for Potable Water Transport

There are millions of kilometers of pipelines that carry potable water (and water for sanitation) from treatment plants to households in cities worldwide. **Pipelines are lacking in many urban centers in less developed and developing countries for the delivery of clean water to millions of people in shanty towns and slums within cities or those on their peripheries.** This is a failing that can be remedied with the construction of treatment plants as might be needed and emplacement of pipelines to municipal wells or to permanent housings in the unserviced urban or edge neighborhoods.

From the perspective of moving safe water to locations where people live with chronic water deficits, one may ask if water can be supplied to them through pipelines from regions with plentiful water supplies, politics aside. For example, in the Middle East, Kuwait could pay to build and maintain a pipeline to bring water from Iraq and reduce its reliance on desalination plants. Similarly, Saudi Arabia and the UAE have the economic resources to build and maintain pipelines to import water from Iran, also reducing their reliance on desalination facilities, if or when political problems are resolved. In Africa, countries with populations that would benefit greatly from the import of water do not have funds to invest in such projects. They would need low cost or no cost loans from international organizations such as the World Bank or from the African Development Bank or grants from development agencies in the European Union, the United States, Japan, and others to do so. For example, in northwest Africa, Burkina Faso could import water via pipeline directly from Mali, especially important for 2050 when the Burkina Faso population is projected to double to 41 million citizens. In southern Africa, Botswana could import water via pipeline from Angola, Zambia, or Zimbabwe. Provider nations would benefit economically from the sale of water and by employment opportunities for their citizens. In all such projects the cost of imported water would have to be set with consideration of local economics and user earnings. The pipeline solution to meet water needs in water poor areas is further elaborated on in Chap. 7 (Sect. 7.6).

2.5.3 Aqueducts

Aqueducts are artificial channels built to carry water from one location where it is plentiful to another where it is needed. They may be comprised of a combination of pipes, tunnels, canals and bridges and in Roman times were gravity fed following a fall of 10 ft (~3.05 m) for every 3200 ft (~975.4 m) of length and mainly underground. They provided potable water, and water for personal hygiene, and irrigation. Modern aqueducts serve the same needs and may be partially gravity fed but when required, use pumping stations to move water upslope as needed in the water moving systems.

Aqueducts graced Rome and its natural environs and areas conquered during the realm of the Empire. They are found in Italy, Germany, France, Spain, Greece, Turkey, North Africa and the Middle East. Ancient Rome was serviced by 11 aqueducts. One built during 19 BC still supplies water for Rome's Trevi Fountain and for two other fountains in the city's center. Another built in about 50 BC in Segovia, Spain still carries fresh water to the city. The Pont du Gard aqueduct/bridge in southern France was build in 19 BC and is the highest and best preserved of Roman aqueducts and a tourist attraction. It is being studied for possible use as an aqueduct in the future. All in all, the Romans built 258 aqueducts totaling a length of 415 km (258 mi).

Aqueducts have been constructed and are in use on all continents to supply drinking water for cities, water for hygiene and sanitation, and water for agriculture (irrigation and animal husbandry). Some transfer water short distances and others long distances. Some are regional in their reach and others are national.

Regionally, the largest aqueduct system serves the mega-city of Los Angeles, USA with the Colorado River Aqueduct (built from 1931 to 1941) moving water from the east 248 mi (400 km) to Los Angeles. The California Aqueduct transfers water from northern California 444 miles (714 km) as the crow flies linearly from Oroville, California (Sacramento-San Joaquin delta) to Los Angeles with diversions to the west and the east that use added aqueduct sections. Construction began 1963 and tie-ins can be added as needed. Other arid western US cities, Phoenix and Tucson, Arizona receive water from the Central Arizona Project that moves water 336 miles (541 km) from Parker, Arizona on the Colorado River. Aqueducts provide water for New York city, another densely populated metropolis. The New Croton Aqueduct built in 1890, the Delaware Aqueduct built in 1945, and the Catskill Aqueduct built in 1960 supply 10%, 50%, and 40%, respectively, of New York City's water needs from the Eastern and Western Catskill Mountains, 90–120 miles (145–193 km) to the north, mainly through tunnels.

Nationally, the Israel National Water Carrier, constructed from 1953 to 1964, is a 130 km (80 mi) long aqueduct that transfers water from the Sea of Galilee (known locally as Lake Kinneret) through pipes, open canals, and tunnels and large scale pumping stations to move water to urban centers and populations in the arid south. The Sea is full with fresh water when its water level is at 208.8 m below sea level. Pumping would be halted if the water level falls to 213 m below sea level to prevent ecological damage. On August 14, 2019, the water level was at 211.48 m below sea level but because of heavy rains in the waterways that fed the Sea during the 2019–2020 winter, it was full in May, 2020.

2.5.4 Other Possibilities to Supply Potable Water to Areas Suffering Severe Water Shortages

Tanker trucks have been used to supply water from where it is plentiful to populations relatively short distances away and road accessible to where there are severe shortages of potable water caused by extended drought periods. This is at best a short term solution to what could be a long term problem such as the 2000s Millennium drought in South/Southeast Australia cited previously.

Are tanker ships a possibility? Another short-term fix for a drought stricken population could be the filling of supertankers used for oil with potable water (after cleaning the hold of oil residue) for transport to water deficit locations. Given sailing time, such a plan has to be instituted before a water poor society becomes "dry". Some have looked to icebergs as a source of water during times of extended drought. At the beginning of the 1900s, Callao, southern Peru, the country's major port, was suffering a severe drought and had a great need for water during the **austral winter**. Improbable though it seems, the government sent a sailing ship to capture an Antarctic iceberg and tow it to Callao. The ship was successful and the need for water for the Callao population was alleviated. A similar plan was proposed for Saudi Arabia, a country always in need of potable water but the route of tow from Greenland or the Arctic would have to pass through warm temperature zones with the result that an iceberg would likely reach port as a greatly reduced mass.

2.6 Some Numbers on Earth's Fresh Water Carrying Capacity for Human Needs

There are 93,120 km^3 **surface sources** of fresh water from lakes, rivers (plus ponds, streams and creeks, reservoir lakes). Desalination adds only ~0.1 km^3 to the earth's clean water inventory. If 70% of the 93,113 km^3 is used for agriculture (food security) and 22% for industrial needs, 8% or 7450 km^3 is available for domestic use. This translates to 7450 trillion liters for drinking, cooking, hand washing, sanitation, personal hygiene including bathing, and laundry that most people use every day, treated and not treated, to serve their daily needs [2]. However, because of the irregular distribution of water globally, a reasonable WHO per person allotment (e.g., 50–75 l daily) from this volume of water is not available to many in the world population. Sadly, there are 1.1 billion people globally in 2020 without access to any type of improved potable water source. This results in 1.6 million deaths annually from diarrhea diseases including cholera with 90% being children under 5 years old. The clean water availability prospects for all in a projected ~9.9 billion global population in 2050 or increased populations later in the century are not favorable unless investments are made to treat and move water via pipeline from where this life sustaining commodity is plentiful to where populations live with deficits that without sufficient water translate to deprivation, illness, and deaths. Pipelines are preferred to aqueducts in many regions (e.g., Middle East, North Africa) because of water loss to evaporation.

Some people walk 1000 m or more to obtain 20 l/day/family member that serves for drinking, hand washing, and basic hygiene. Others have immediate access (<100 m) to 50 l of water/family member/day that adds bathing and laundry to its use. However, reality emphasizes the disparities in safe water availability. The earlier cited UN estimates that by 2050, ~4.2 billion people (of a projected ~9.9 billion) will be living in countries that can not supply the 50 l/person/day [4]. The UN further reports that in 2019, 14 African countries face water scarcity and estimated that by 2025, nearly 1 in 2 Africans will be in areas of water scarcity or water stress, mainly because of population growth [17]. To emphasize the problem today and realize what water problems will arise in Africa in coming decades to 2050 or later in the century, it is important to know the time used to collect water at the expense of other productive efforts. In 25 Sub-Saharan nations in recent years, women spend an average 16 million hours a day, men 6 million hours a day, and children 4 million hours a day collecting life sustaining water.

Satellite analyses for Africa used the Tropical Rainfall Measuring Mission to estimate rainfall and the Gravity Recovery and Climate Experiment used changes in the earth's gravitational field to infer changes in the global water resources from surface and aquifer waters. Analyses of the satellite data suggested that UN estimates do not account for deep water aquifers to give total water availability [18]. However, these analyses could not determine if the deep water aquifers inferred contained fresh or brackish water and whether they were unconfined or more likely confined. Nonetheless, it is probable that even deep aquifer water availability could not solve African nations water deficits in the future (e.g., by 2050) because of the high rate of population growth especially in Sub-Saharan nations (1,094 billion in 2020 to 1,591 billion in 2035 to 2,192 billion in 2050).

According to the WHO, 75 l/person/day would protect against disease: 3 l for drinking, 4 l for cooking, 20 l for bathing, 40 l for sanitation, 22 l for washing vessels, and 23 l for gardening. A recent study estimated that to service all their needs (e.g., food, water, sanitation, goods, etc.), a per person requirement is 500,000 l annually or 1370 l daily. Eight percent for domestic daily needs would be ~110 l [18]. This corresponds well with other estimates that 100 l/person/day of clean water at a tap would meet all consumption and hygiene needs but was realistic in stating that in rural areas only 30–40 l/person/day was accessible and that in remote villages as little as 4 l/person/day was obtainable for domestic use. This, vs. 380 l/person/day used in the United States or 314 l/person/day in Japan illustrates the great disparity in water availability/use between water-advantaged and water-disadvantaged populations. Regionally, the average per capita daily domestic use in

North America is **428** l, in Europe **280** l, in Asia **132** l, and in Africa **63** l. The global water inventory has the carrying capacity to support a good quality of human life, but sadly it is not geographically accessible to all in what might seem to be an unattainable ideal world condition for populations by 2050 or later in the century. Only pipelines to carry water can help develop a greater carrying capacity of this life sustaining fluid.

Afterword

In a future chapter, we will discuss how global warming/climate change will alter the water picture and its effect on growing global populations and demographic changes projected to take place as the world reaches 2050. What regions will be most affected? What regions will be little affected, if any? What will be water availability effects on humans and other life forms? Can we expect water refugees? Could need for water result in water wars?

References

1. Shiklomanov, I.A., 1993. World's fresh water resources. In Gleick, P., Water in crisis: a guide to world's freshwater resources. Chapter 2, pp. 13-24. Oxford Univ. Press, New York
2. United States Geological Survey, 2019. Directive. How Much Water Is There On Earth? Online. www.usgs.gov/.../science-how-much-water-there-earth
3. UNFAO. Aquastat, averaged from 2013–2017. For values of renewable water resources www.fao.org/water/aquastat/data/query/index.html
4. Moncrieffe, J., Lead author/researcher, 2008. State of the World Population 2008. United Nations Fund for Population Activities, 108 p., New York
5. Population Reference Bureau The World Population Data Sheet 2019. www.prb.org
6. www.fao.org/water/aquastat/data/query/index.html
7. Hofste, R.W., Reig, P. and Schleifer, L., 2019. 17 Countries, Home to One-Quarter of the World's Population, Face Extremely High Water Stress. World Resources Institute. Online. www.wri.org/blog/2019/08/17-countries-home-one-…
8. Wang, T. 2019. Desalination capacity of operational plants worldwide as of 2018, by region (in million cubic meters per day). Online. www.statista.com/statistics/960259/capacity…
9. Folk, E., 2019. Environmental Impacts of Seawater Desalination. EcoMena. Online. www.ecomena.org/environmental-impacts-of-…
10. Jones, E., Qadir, M., van Vliet, M.T.H., Smakhtin, V. and Kang, S., 2019. The state of desalination and brine production: A global outlook. Sci. Total Environment, 657: 1343-1356.
11. Gies, E., 2019. Slaking the World's Thirst with Seawater Dumps Toxic Brine in Oceans. Scientific American, February. Unpaginated. Online. www.scientificamerican.com/article/slaking-the-thirst..
12. Gies, E., 2019. Desalination Breakthrough: Saving the Sea from Salt. Scientific American, June. Unpaginated. Online. www.scientificamerican.com/article/desalination-breakthrough…
13. UNESCO, 2017. Wastewater: The Untapped Resource. UN World Water Development Report. Online. www.unesco.org/.../wastewater-the-untapped-resource

14. International Desalination Association, 2019. Dynamic growth for desalination and water reuse in 2019. Unpaginated. Online idadesal.org/dynamic-growth-for-desalination-and-water-reuse-in-2019
15. Wikipedia. Pipelines. Online. en.wikipedia.org/wiki/Pipeline_transport
16. https://en.wikipedia.org/wiki/List_of_countries_by_total_wealth
17. Ekins, P., Gupta, J. and Boileau, P., (Eds.), 2019. Global Environment Outlook 6: Healthy Planet, Healthy People. United Nations Environmental Programme, NY, 687 p. Online. www.unenviroment.org/resources/global…
18. Hasan, E. and Tarhule, A., 2019. We use satellites to measure water scarcity. Unpaginated. Online. theconversation.com/we-use-satellites-to-measure-water-scarcity

Chapter 3
Food Security/Insecurity, Food Systems

3.1 Introduction

Food security has been defined as a situation that exists when all people, at all times, have physical and economic access to sufficient, safe, and nutritious food that meets their dietary needs and food preferences for an active and healthy life [1]. This is not the case in 2020 for large numbers of the Earth's populations especially in those less developed and developing nations, where for some, chronic food insecurity is a major problem.

3.1.1 Food Insecurity

Eight hundred and twenty-one million people or one of nine of the ~7.8 billion on Earth in 2020 suffer from food insecurity. This food insecurity is categorized as being moderate or severe. As defined by the United Nations Food and Agriculture organization (FAO), **moderate food insecurity** occurs when there is uncertainty about obtaining food such that people have to sometimes reduce the quantity and nutritional value of the food they eat. This may be because they are unable to pay for food or provide goods or services in exchange for food. This irregularity in availability of food, often times of poor quality, affects citizens nutrient intake and this in turn can harm their biological health and physical and mental conditions. Under **severe food insecurity**, people may run out of food, suffer hunger, ofttimes not eating for days. This puts their above cited health and physical and mental conditions at great risk that can lead to death (e.g., starvation). There are a larger number of people suffering moderate food insecurity than those suffering from severe food insecurity [2]. As a result of undernutrition, there are 151 million children under 5 years of age shorter (stunted) than their well nourished peers and at risk of cognitive disability. In addition, 613 million women and girls between the ages of 15 and 49 (child bearing ages) suffer from anaemia (iron nutrient deficiency).

This should not be the case because **enough food is being produced globally in 2020 to supply nutritious food to all the Earth's human populations,** even with the amount of food that is wasted or lost in food systems (see Sect. 3.4). Nonetheless, undernutrition, malnutrition, and starvation are afflictions that affect many on Earth for several reasons. These include conflicts/wars that prevent delivery of food from where it is produced to where it is most needed, militants that hijack food shipment for their own use or to sell, and governments led by despots that do not permit delivery of food to nutrition starved segments of their populations while they and their allied populations eat well. Then there is the poverty condition when citizens do not have access to available food because of lack of funds to purchase enough nutritious food to keep them healthy and there is no government safety net to provide for their basic needs. Lastly, problems in the farm to table food system infrastructure causes food loss and waste that when 'repaired' could feed food stressed populations. A following section (Sect. 3.4) will discuss the food system in some detail.

Those regions most vulnerable to food insecurity in 2020, Sub-Saharan Africa, South Asia, and South America and the Caribbean, are likely to be more vulnerable in 2050 and later in the century. An interactive map program projects how nations of the world in 2020 and in 2050s and 2080s might fare with respect to food insecurity. The program allows the user to visualize how different combinations of amounts of greenhouse gas (GHG) emissions from a nation (low, medium, or high) and adaptations to combat them (high, low, none) could affect food security. The projections are the result of averaging outputs from 12 CMIP5 models (Coupled Model Intercomparison Project Phase 5) that are based on three components: (1) exposure to **climate** related hazards (e.g., heat, drought, floods); (2) sensitivity of national agricultural production to **climate** related hazards; and (3) adaptive capacity or how a nation would cope with a **climate** related food shock [3]. This includes input on population increases in 2050 and demographic shifts with ~70% of the global population in urban centers (6.9 billion) versus ~55% (~4.3 billion) in 2020. On a smaller physical map scale, one can review the various combinations plus ensemble representations of the degree of exposure of the world's nations to food insecurity with climate change under the varying combinations of conditions [4]. You, the reader, should access and view these maps for an increased perspective on future global food security.

3.2 Soil: A Foundation of Food Security

The FAO defines **arable** land as land under temporary crops (double or triple cropped areas are counted once), temporary meadows for mowing or for pasture, land under market or kitchen gardens, and land temporarily fallow, but with land abandoned as a result of shifting cultivation excluded. Food production and security for the Earth's population depends primarily on water, exposure to sunlight, and

3.2 Soil: A Foundation of Food Security

Table 3.1 Macro- and Micro-Nutrients that are essential to plant/crop yield and nutrition [5, Modified]

Macro Element	Soil concentration [a]mg/kg (= ppm)	Micro Element	Soil concentration mg/kg (= ppm)
Nitrogen	15,000	Chlorine	100
Potassium	10,000	Iron	100
Calcium	5000	Manganese	50
Magnesium	2000	Boron	20
Phosphorus	2000	Zinc	20
Sulfur	1000	Copper	6
		Molybdenum	0.1
		Ni	0.1

[a]mg/kg = milligrams per kilogram or ppm = parts per million 10,000 mg/kg or ppm = 1%

Table 3.2 Examples of relative yield and nutritional response of selected crops to four micronutrients given in Table 3.1 [6, Modified]

Crop	Boron	Copper	Manganese	Zinc
Wheat	Low	High	High	Low
Corn	Medium	Low	Low	High
Soybean	Low	Medium	High	Medium
Alfalfa	High	Medium	Low	Medium
Grain Sorghum	Low	Medium	Medium	High
Oat	Low	Medium	High	Medium
Irish potato	Low	Medium	Medium	High
Tomato	High	High	Medium	Medium
Apples	High	Medium	Low	Medium
Citrus	Medium	High	Medium	Medium
Cotton	High	Low	High	Medium
Rice[a] [7]	Medium	High	Medium	High

Cu + Zn = highest yield, most nutritious rice

extent of **arable soils** and their contents of essential macro- and micronutrients that are required for healthy plant/crop growth (Table 3.1). A concern for agriculturists is the ability to control weeds and pests that can rob soils of nutrients and consume or otherwise ruin crops whether in a soil, on a soil, from bushes or from trees. Soils produce the feed/fodder for land food animals. We will discuss the factors that affect food production in 2020 later in this chapter and will examine the effect of populations' growth and global warming/climate change on them for 2050/2100 in Chap. 8.

Each agricultural crop requires specific micro-nutrient(s) to maximize yield and nutritional value. Selected examples are shown in Table 3.2. Because a soil's nutrient content is diminished seasonally, it has to be replenished and this is done by applying one or more specific micro-nutrients with fertilizer.

There is no doubt that factually, we have the carrying capacity to feed all of the ~7.8 billion people on earth in 2020. In reality, as cited earlier, more than 800 million to 1 billion people on Earth are suffering from undernourishment, malnourishment, and in some cases starvation. As previously noted, the afflicted are often children whose growth from malnourishment is stunted and cognitive ability impaired. Sociologically, part of the malnourishment problem is the result of poverty and a country not having a 'safety net' for the hungry including tens to hundreds of millions of young children and aged citizens that can not fend for themselves. Part of the problem may be the result of climate events that damage crops and kill food animals. This issue is exacerbated in conflict areas by factions that steal food during transport to the hungry to feed their followers or in order to sell it. Some governments may use the withholding of donated food as a political/controlling tool against opposition segments of their populations [8].

3.2.1 Minimizing Soil Erosion

Food production/security sustains life. Given enough precipitation to support rainfed crops (86% of total plantings) and water from rivers or aquifers for cropland irrigation (16% of cultivars), and enough exposure to sun, fertile soils will sustain food production for human populations as well as for food animals (fish excluded) that contribute to food security.

Priority then for food production and hence security is the preservation of a fertile soil mass from degradation. The principal degrading factor is erosion by moving (flood) water and/or wind. Various methods are used to minimize soil erosion of agricultural fields. First and foremost is to avoid exposed soil surfaces. This is done by reducing tillage that breaks up a soil when preparing it for planting by using equipment that slices into a soil and seeds it without digging up clods of soil that have great surface areas exposed to the erosive force of wind or moving water. Such equipment is not readily available to farmers in less developed and developing countries because of its costs that cannot be borne by small farm agriculturalists although it would limit soil degradation by erosion. Government investment in this equipment is warranted if purchased with the mandate that it is to be shared among small farms that account for much of the food crops in many regions. Wind erosion is minimized by windbreaks around agricultural fields. These are 'green walls' of native trees or bushes that have minimum water needs. If eroding winds are seasonal and flooding is not a threat and water security is good, wetting the soils can reduce wind erosion. Another factor that can create a soil erosion scenario is overgrazing grasslands (by cattle, sheep, goats) so that grazing management should be a control on this potential problem. Active management programs in areas poorly served in the past in many less developed and developing countries now have outreach programs. Agents work with farmers and food animal producers to put into place methods to best counter erosion and grazing problems so as to protect their food production capabilities for themselves and for markets, hence their livelihoods.

In the future, perennial grains will be useful erosion control plantings. Perennial grain is one that remains productive for 2 or more years so that ground does not have to be prepared for seeding annually, thus decreasing annual exposure to erosion. One such grain has been under development since the 1980s and has recently been put into production. This is an intermediate wheatgrass (Kernza®) for which one product has been manufactured: Long Root Ale. Research on other grains that may lend themselves to perennial growth cycles is ongoing. Kernza® kernels are being used in a few restaurants for pizza dough, pastries, salads, pancake batter, and flat bread crackers. Because the (Kernza®) kernel is small and bushels yield per acre is low (compared to annual wheatgrass) more research and development is needed before significant commercialization in high volume market cereals is possible [9].

3.2.2 Maintaining Soil Nutrient Contents

To sustain productivity of a soil, and thus food security, macro- and micro-nutrient contents have to be maintained (Table 3.1). Nutrient content is continually extracted from soil to support crop yield and nutritional value. A soil nutrient content can be replenished as necessary during cropping cycles two ways. One is favored by organic farmers and involves growing green manure and cover crops (e.g., winter rye, winter wheat, crimson clover, hairy vetch) seeded in late summer or early fall. These crops may also add protection from erosion during the winter months. Some crops may be killed by the cold and decompose in place delivering nutrients to the soil. Others are sown back into a soil before they flower in early spring 2–3 weeks before planting summer crops. This enriches a soil in nitrogen and other nutrients. The other, used by most commercial farms, is by applying (agricultural) chemical fertilizers as needed to replenish a soil with nitrogen plus other nutrients and micro-nutrients that are critical to the growth of a particular crop such as those listed in Table 3.2.

The practice of conservation agriculture with no or little mechanical soil breakage, with the use of organic mulch soil cover, with crop species diversification plus other good management methods previously cited is expanding in use. Conservation agriculture has grown globally almost exponentially from 2.8 million hectares (ha) worldwide in 1973/1974 to 72 million ha in 2003 to 106 million ha in 2008/2009 to 157 million ha in 2013 or ~11% of the ~1427 million ha farmed globally. The United States, Brazil, Argentina, Canada, and Australia account for 132 of the 157 million ha under conservation agriculture [10].

It is clear that sustainable management that minimizes erosion of farmland and assures nutrient replenishment of soils is essential to protect food security. Food security is being further enhanced by land restoration and rehabilitation of open fields for increased agricultural production. Investment in such projects has, in some cases, given a benefit to cost ratio of 3–6 times to 1. The cost to bring what was a productive fertile soil back to good food production and improved food security varies greatly from US$20 to US$5000 per hectare (2.5 acres) with a median of $US500. Such land restoration and rehabilitation should be modus operandi for degraded farmland globally where future benefits are greater than existing costs to do so.

Food security can be further enhanced beyond 2020 when erosion and nutrient replenishment are controlled by farmers. This is done with the use of seeds for cropping local food plants that have been genetically modified to be tolerant of heat or drought and resistant to chemical control of weeds, pests, and diseases. The seeds may also be altered to improve crop yield and nutrition. Improved fodder crops and genetically modified food animal breeds can minimize the use of anabolic steroids (growth hormones) and whole herd inoculations against disease used in some high volume/high density commercial operations (cattle feedlots, dairy farms, and poultry farms).

3.3 Population Challenge to Food Security

It should be emphasized that food security is being challenged not only by the need to be able to feed more people under food stress today and in the future, but also by economic status and changing dietary habits. Within the Earth's growing populations, greater numbers have been rising out of poverty worldwide. With higher incomes there are more consumers with greater demands for food products previously not financially accessible to them. For example, the demand for meat since 1961 has increased more than twofold and caloric intake has increased by 1/3. The consensus of food scientists estimates that food production (for humans and food animals) should increase gradually year by year to reach a 70% higher production to have the food capacity to carry the projected additional population in 2050 [11]. Thus, for example, an annual cereal/grain yield should increase to 3 billion tons by 2050. Concurrently, meat will need to increase to 470 million tons by 2050. A 2013 report estimated the need is to double food production to sustain food security for 2050 populations [12]. Still another report in 2017 sets the need to be 25–70% greater crop output by 2050 with the caveat (improbable) that nutrient losses and CO_2 emissions drop dramatically so as to rejuvenate and maintain ecosystem viability [13]. Likewise there will be more demand to grow food animal feed.

In Chap. 8, the discussion will continue on how food production can be increased to provide food security and carrying capacity for the 2050 population projected to grow by 2.3 billion persons in 30 years. By 2050, the number of people living in cities and not producing food but rather shopping for food in markets will grow from 4.3 billion in 2020 to 6.9 billion. Can food production and food transport and delivery be ramped up and be sustainable over decades to meet population needs especially for non-food producing inhabitants of highly and densely populated urban centers? This, within the context of more people with more disposable income to foster high protein, high caloric dietary changes on the one hand, against hundreds of millions at the poverty level less able to compete for enough nutritious food. The failure of any national leaders to prioritize planning ahead to assure sustainable food availability and access for all citizens as 2050 approaches and global populations expand is a grand mistake. It risks food riots and calls for changes in government.

3.4 Food Security Dependency on Food Systems

Food security depends on a smoothly (as possible) functioning food system that starts with the production of food and ends in the delivery and consumption of sufficient and nutritious food that sustains good health. Food systems are multifaceted and may start with the sowing of seeds and cultivation of crops, breeding and growth of livestock, and conservation of fisheries in all aqueous environments. Upon maturity, these food sources are harvested and/or prepared in a production phase and stored before transport (via road, railway, barge/boat, air) to processing facilities where they may be packaged or otherwise prepared before being stored again until delivered to retail markets and restaurants. Food systems finish their cycles with consumption. In all phases of a cycle, food is often lost or wasted at markets, homes, and food establishments [14]. Farmer education and the use of improved technology can reduce agricultural food losses by 90–100%. For example, grain is lost post harvesting to spoilage by rodents and bacteria during container storage but could be protected from such loss using hermetic storage containers [15]. Stored food at wholesale and retail locations that can be lost to warm conditions, such as fresh fruits and vegetables, can be protected using or expanding cold storage facilities. Food loss during transport in warm seasons (or warm environments) can be reduced by refrigerated transport. This will require directed investment by food system businesses to sustain profit, perhaps subsidized to some degree by nations that want to maintain and improve food security as populations increase. The food waste at markets, homes, and food establishments can be reduced as mandated in France by cycling edible comestibles to food banks for distribution to needy people (see Sect. 3.4). The reduction of food loss and food waste is contributing to feeding needy populations in 2020. As the amount of the reduction of food loss/waste continues to increase, the saved food will serve in need populations as exist and likely grow in the coming decades of the twenty-first century.

Food systems and hence food security are under constant pressure from several interrelated stressors that have already been mentioned. These include the need to feed growing global populations and demands for more and better quality foodstuffs from greater numbers of well employed citizens with disposable income that have emerged from poverty and entered the middle classes in less developed and developing countries. During the past few decades, this has resulted a growing demand for basic food crops (e.g., maize, wheat, rice, vegetables, fruits) as well as for livestock (meat) products. Both crop and animal husbandry production are subject to decreased output temporally from disease (e.g., swine flu and pork availability) and climate change events in some regions such as increasing temperature, changing precipitation patterns, and intensity, frequency, and duration of events that negatively affect farm production (e.g., heat waves, drought, flooding). For example, reports raise the probability that yields of some crops (e.g., maize, wheat, barley) will decline with rising temperature in high population growth low latitude regions. However, at higher latitudes (slow population growth or contracting populations), increasing temperatures in coming decades will increase yields of some crops (maize, wheat, sugar beets, cotton). [1, 16]

3.5 Food Production: The Basis for Food Security

Production of comestibles can be improved for existing nutrition poor populations in three ways and provide as well for growing populations especially in less developed and developing nations in South Asia and Sub-Saharan Africa. One is to use seeds tailored to the growing environment by traditional hybridization or by genetic modification to give good yielding and nutrition sufficient crops, Another is to adopt crop management practices (e.g., for soil quality, irrigation, and fertilizer, herbicide, pesticide application) and improved animal husbandry methods (e.g., grazing land management). A third that is not encouraged but that may be necessary is to expand the hectares devoted to food production.

The FAO estimates that food production will have to grow by ~50% on 90-to-325 million more hectares of cropland to meet the goals of providing enough food with acceptable nutritional value for the projected 9.9 billion 2050 population. This estimate depends on climate change and how much loss of food and food waste is reduced or demand lessened by citizens' dietary change to consumption of less or more livestock [17]. Another report is based on earth conditions in 2050 if nothing is done to substantially to reduce global warming with 650 ppm CO_2 in the atmosphere and a 2.5 °C temperature increase over pre-industrial temperature (IPCC RCP 8.5 condition). It suggests that 100 million ha of additional farmland would be needed to double food production to sustain global populations, again especially in Sub-Saharan Africa, South Asia, and to some degree in South America [18]. To date, there are computer crunched numbers based on factual measurements and observations but no fixed answers to food/acreage needs. This is because of changing input data for variables to computer systems that affect predictions for coming decades (see Sect. 3.1.1, third paragraph). That the Earth's food carrying capacity to feed a growing population will have to increase as the twenty-first century decades pass is a given but a given with more than one pathway that could achieve that end.

Of the more than 13.4 billion ha of land surface on our planet only 3 billion ha are suitable for crop production with half now under production [19]. Use of any land from the arable remaining hectares has to be assessed carefully in the effort to minimize threats to the biological diversity that supports ecosystems and their usefulness for sustaining societal needs.

3.5.1 Increasing Food Production

Increasing food production that supports a well functioning food system can be approached in different ways by agriculturists on existing or expanded cropland. One way already noted is to mitigate or essentially stop soil erosion and a second is by applying fertilizer where soils develop nutrient deficiencies. Deficiencies are revealed by seasonal chemical analyses or when farmers note a drop in crop yields. The IPCC reported that since 1961, an **8×** increase in nitrogen fertilizer use coupled with a doubling of crops under irrigation improved the per capita food availability in 2018 by ~30% although the global population increased **~2.5×** from ~3.1 billion to 7.6 billion people [1].

Crop loss may be reduced by the use of herbicides and/or pesticides and other methods that minimize the impact of disease on yield. However, although these chemicals may increase crop yield, they, like fertilizers, present an environmental tradeoff. They should not be used in excess because runoff from fields can be dangerous to natural resources such as surface and groundwater, and lifeforms such as fish. **Agricultural chemicals overuse is a waste of money for farmers.** The danger to the environment from the use of agricultural chemicals (less those for nutrient replenishment) can be avoided in some areas with seeds modified by traditional or marker-assisted hybridization or by genetic engineering. Modified seeds can reduce the use of agricultural chemicals to control weeds, pests, and disease as well as for improving crop yield and nutrient content. For example, some genetically modified rice and corn species have had higher yields and improved nutritional values, a benefit for consumers [20].

3.5.2 Speed Breeding by Extended Photoperiods Exposure Adjunct to Hybridization

In a speed breeding program, researchers have subjected immature seeds to extended photoperiods in controlled glasshouse environments to accelerate growth of food crops in order to study successive generations for favorable crop traits such as resistance to drought and increased yield. Field growth of such plants in traditional hybridization programs is slow, yielding 1–2 generations yearly. Under extended periods of 22 h of light under high pressure sodium LED lamps and 2 h of dark, researchers were able to induce seed germination to yield 4–6 generations of the food crops spring wheat, durum wheat, barley, chickpea and pea plants in a year and 4 generations of canola per year vs. 3–4 in field growth conditions. This speed breeding technique accelerates research on phenotyping of adult plant traits and plant mutations for genomic studies. Added to other cutting edge food crop studies, this speeds up identification of plants with traits that can be imparted to seeds that give crops higher yields, improved nutritional values, and resistance to short-term and long-term aberrations in environmental conditions (e.g., temporary submergence, drought). This can enhance food security for the growing global population, especially in vulnerable less developed and developing regions [21].

3.5.3 Mutation by Space Exposure to Cosmic Radiation

In 2006 China sent 200 kg of fruit, vegetable, grain, and cotton seeds into space in an experiment that exposed the seeds to cosmic (ionizing) radiation in zero gravity. The Chinese **claim** that the planting of the radiation exposed **grain seeds** is reported to have resulted in a yield increase **11% higher** than the traditional variety in China's annual grain production. If this gives the same result on a commercial scale, gives better results than hybrids or genetically modified seeds, and can be reproduced with human generated ionizing radiation, it would benefit China's food production but this has yet to be tested and reported on. Subsequent Chinese space probes bearing seeds of crops and flowers have been returned to Earth but publications (in English) on experimental results are difficult to find. Studies on plant and insect mutation by exposure to cosmic radiation is not new and has been carried out over the years on the International Space Station. Individual countries with access to space probes have similar programs but at a smaller scale because the costs of such dedicated programs is deemed too costly. However, the Chinese government considers the investment worthwhile and expects to reduce the costs of space experiments and use the results to increase its citizens food security [22].

3.5.4 Expansion of Farmland

Lastly, food production can be increased by expansion of hectares for crops and animal husbandry. As cited previously, about 1.5 billion ha (15 million km^2) of land surface are amenable for expansion (see Sect. 3.5, third paragraph).

Of >23 million km^2 of high biodiversity areas that have been identified by the Alliance for Zero Extinction, 900,000 km^2 (90 million ha) or 3.76% are projected for cropland expansion by 2050. This represents 6% of the 1.5 billion ha that could be opened for cultivation. However, these areas house >50% of the world's endemic plant species and 77% of terrestrial vertebrates. With this in mind, one can question the cost to vulnerable biodiversity and by extension to people dependent on life forms for other than direct food availability (e.g., pollination), and direct food production (e.g., soil biota interact with plant roots and contribute to nutrient distribution as part of the soil food web) [23]. To balance population requirements against losses to societies, land use management and environmental legislation is necessary to prevent cropland expansion into specific areas. This includes forested and other vegetated terrain with habitats that sustain fauna and flora and the natural resources they provide to people [19].

In a study that emphasizes biodiversity preservation while striving to increase food production, researchers evaluated four recent global maps of human influence. They estimated that the Earth's terrestrial surface, exclusive of permanent snow or ice-covered land has areas (ecosystems) with **very low** (20–34%) and **low** (48–56%) **human influence** [24]. Many of these areas are in cold regions (boreal forests, montane grasslands, and tundra) or arid zones (deserts). The areas with **low to very low human influence** with the ecosystems and the biodiversity they contain should be given high priority for preservation and resources they may provide for the tropical and temperate areas with very high human influence >99% as shown in most map data sets [24]. As discussed in the following section, some countries have prioritized preservation of biodiversity in very low and low areas of human influence above using them for cropping and raising food animals.

3.5.5 Population Growth and Farmland Expansion

Tropical Asia is comprised of 16 countries with a population that has grown from ~1.84 billion people in 1999 to ~2.43 billion in 2019, an increase of close to 25%. To feed the additional almost 600 million citizens, the region suffered more deforestation and faster opening of land for farming than any other region during that 20 year time frame. This was at the expense of the largest percent of biodiversity lost to cropland. In addition to food production, further deforestation in the region supported timbering and non-food producing plantations that yielded export products such as wood and palm oil that favored business and added to a nation's GDP but at a cost to future direct and indirect societal benefits from healthy, intact ecosystems.

In the two decades from 1999 to 2019, the population of the 48 countries that comprise Sub-Saharan Africa increased by ~70%, from ~630 million to ~1.07 billion people. By 2035, the projected Sub-Saharan population of 1.57 billion people (~47% increase over 2019) will be greater than that of China and close to the projected 1.58 billion population in India. To meet the need to feed the rapidly growing Sub-Saharan populations will require a dramatic cropland expansion into biodiversity rich regions in the near future (2035) and beyond. This will threaten the habitats of a great number of animal and plant species. However, with foresight in planning for agricultural development and with continued monitoring and management of land use of Sub-Saharan terrain, the negative impact on biodiversity can be significantly lessened.

Latin America is a region with a promising inventory of agricultural land for expansion, especially in Brazil and Argentina. Population growth in Latin America and the Caribbean region was less than Tropical Asia and Sub-Saharan Africa. The region's population in 1999 (519.2 million people) increased ~20% by 2019 (660.3 million people). The 2035 population is predicted to grow to 742.7 million people, far less proportionally than the previous cited two regions' projections. Latin America has accounted for ~16% of total global food and agricultural exports (e.g., bananas, beef, coffee, corn, poultry, soybeans, sugar) [25]. This percentage can be increased. In order to protect ecosystems in Latin America as in Tropical Asia and Sub-Saharan Africa, and what they impart to human health and quality of life, national legislation should define areas protected against cropland expansion and be rigidly enforced. This is especially important for Brazil because of the country's high rate of deforestation to satisfy agribusiness and national development goals (similar to Indonesia). Deforestation should not be allowed to follow the whims of economic-political decisions at the expense of biodiversity. Proven corruption of government decision makers that bypasses laws and contributes to deforestation or other uses of terrain that detracts from its usefulness to societies into the future should be punished with heavy fines and/or incarceration.

There have been suggestions that the need for agricultural expansion can be lessened by changing consumption patterns to less animal-sourced and more plant-based diets (discussed in Chap. 8). Although true, the possibility that this can make a substantial difference is limited because of inculcated consumption habits and increasing incomes of citizens rising out of poverty and entering a growing middle class with a degree of expendable income to spend on heretofore available but financially inaccessible foods. In addition, the constant flow of food commercials on television and other popular media can attract those who can spend money on animal-sourced foodstuffs [26].

3.5.6 Improve Food Security: Reduce Food Loss, Food Waste

Currently, food security studies of food systems estimate that one-third of food produced globally is wasted or lost to poor storage and transport conditions. Of the quantities produced, most wastage is for fruits and vegetable plus roots and tubers (40–50%), 30% for cereals, 20% for oil seeds, meat, and dairy, and 35% for fish. This amounts to 1.3 billion tons of unused edible food. In theory, if just 25% of this is recovered, it could feed 870 million people that in 2020 could end world hunger that affect the more than 810 million people that suffer from undernutrition, malnutrition and starvation [11].

Annual per capita food production for populations in rich countries is ~900 kg (~1900 lb) per capita, a mass that is almost double the 460 kg (~1012 lb) per capita annual production in the poorest regions. Overall, rich country consumers **waste** 222 million tonnes of edible foods, an amount almost equal to the **net food production** of 230 million tonnes in Sub-Saharan Africa. For example, the annual per capita waste in Europe and North America is 95–115 kg (209–253 lb) vs. 6–11 kg (13.2–24.2 lb) per capita annually in Africa and East and Southeast Asia [27].

The causes of loss and waste of edible foods vary according to economic conditions. For example, developing countries have 40% loss in harvesting and processing whereas developed countries have 40% loss at retail and consumer levels. Reduction of food loss and waste can be accomplished in all of the food system. To do this, countries can improve harvesting and storage of foodstuffs, improve the infrastructure that supports food production, the transport type and timing, the packaging and retailing with honest labelling with a **'best used by' date and 'suitable for use'** as a second date. In the United States, only baby food has a best used by date that has to be honored. Other foodstuffs carry a manufacturer's dates that do not reflect the extended dates foods are perfectly edible. Manufacturers' dates may scare consumers into throwing away perfectly good food.

France passed a law prohibiting the disposal of edible foods from markets punishable by large fines (€3750). Of the 7.1 million metric tons of wasted edible food generated in France, (234 lb [106 kg] per capita annually), 11% (781 thousand tonnes) was at supermarkets. This edible food is now collected and distributed to those who need nutritious food to ward off hunger and keep healthy. Although restaurants in France are not bound by the law, their managements have joined the effort to collect and distribute edible food to the needy. Fortunately, needy populations in many cities globally are benefitting where edible nutritious foods are voluntarily collected from markets and restaurants and distributed to citizens following the French example.

Unnecessary waste in the food system is a real economic loss as illustrated by those at the farm level alone. Lost is the cost of the water used for crops and to hydrate food animals, the cost of added soil nutrients, the cost of herbicides and pesticides applied, the cost of human labor used in seeding, care, and harvesting, and the cost of energy used. Together with other cost losses in the food system, the result is a major drain of investment capital. The financial debit from the loss and waste of food in industrialized countries has been estimated as US$680 billion (670 million tons) vs. US$310 billion (630 million tons) in developing countries. The losses are considered as the cost of doing business and are made up by higher food prices.

Edible food disposed of in landfills can be detrimental to the environment. As organic matter decomposes, it generates methane, a greenhouse gas that contributes to global warming as it leaks in to the atmosphere. This adds to the methane gas that emits from ruminants (food animals) and rice farming. These latter sources contribute to food security but are responsible for 1/2 of agricultural source of methane. Scientists are researching how the volume of methane gas from ruminants can be reduced via changing their feedstock. For example, recent **laboratory experiments** in Australia showed that dried seaweed (*Asparagopsis taxiformis*, a red macro algae) in rumen food reduced methane output by 99% [28]. Subsequent **field studies** showed that when the dried seaweed made up 2% of sheep diets during a 72 day period, their methane production was 50–70% less. Given that 10% of GHG emissions in Australia is from ruminants, this represents a good step forward and stimulates research into how 'artificial' seaweed can be produced to serve the demands for the product without disrupting a marine ecosystem. Other researchers are studying dietary changes that will reduce methane emissions from flatulent activity in beef cattle and dairy cows.

3.6 Adding Farmland from Within for Food Production

The total global land area has a 2018 extent of 127.3 million km² (~12.7 billion ha). Rural land area represented 111.8 million km² (11.2 billion ha). As previously noted, 3 billion ha of the rural land area are suitable for crop production with half now under cultivation of crops [19]. In addition, use of any land from the remaining cultivatable hectares has to be assessed carefully in an effort to minimize threats to the biological diversity that enhances sources of future food security and supports vital ecosystems. Globally, the best open areas for agricultural land availability and development for additional cultivation to expand food security are in Africa and South America.

From 1961 to 2016 the agricultural area per capita worldwide in hectares declined from 0.367 to 0.192 [29]. This number will shrink further as populations grow and there is competition for more land to accommodate living/working space for citizens and for land to produce crops for their food, and fodder and grazing land for food animals, especially in Africa and Asia. The solution to this problem that is being followed is for urban centers is to expand upward rather than laterally leaving cropland intact. Indeed, to sustain food requirements for growing populations, especially in cities, some new areas will have to come under cultivation and food animal production. This will protect food security even as hybridization and genetic alteration of crop seeds render greater yields of more nutritious crops. In addition, any thought to convert grasslands and forested areas to cropland should be carefully evaluated vs. the benefits of preserving these vegetated areas for grazing acreage and fiber and timber resources and ecosystems that support biodiversity and sinks for CO_2 emitted to the atmosphere, the principal driver of global warming.

3.6.1 Reclaim Cropland from Fuel Feed Stock

In order to reduce CO_2 emissions from the use of 100% gasoline in vehicles, a 10% ethanol 90% gasoline mixture is used internal combustion engines. This use for ethanol increased market demand that resulted in the conversion of food and feedstock acreage to crops that supported ethanol production especially when subsidized by governments (e.g., the US). In some countries, this reduced the acreage for food production/food security. Of the ~86.4 million acres planted for corn (maize) in the United States during 2018, ~11% or ~9.5 million acres was for the production ethanol from the starch in the kernels. In Brazil, vehicles use 100% ethanol produced from sugar cane [30]. In addition to reducing CO_2 emissions, the ethanol additive also reduces emissions of particulate matter, and oxides of nitrogen, contributors to smog formation. The United States is the largest global producer of ethanol (58.8%) followed by Brazil (27.9%). An increased global use of hybrid and electric vehicles in coming decades as prescribed by law in many nations will weaken the demand for ethanol. General Motors announced in 2021 that by 2035, it will produce only EVs. Other companies will follow suit but perhaps not on the same schedule. A result will be that millions of cropland acreage devoted to ethanol production will be available to return to food and feedstock production. This will globally augment the Earth's food carrying capacity for growing populations.

3.6.2 Planting GMO Crops Can Free Land for Additional Food Production

Genetically engineered (GE) crops can significantly reduce the use of agricultural land per crop because of **higher productivity** (yield and nutritional value) and herbicide tolerance and/or pest/disease resistance. Thus, expanded plantings of genetically engineered (GE) crops globally will free up cropland for additional food production. For example, during a 21 year period (1996–2016), reports show that GE maize had 5.6–24.5% higher yields and enhanced grain quality compared with non-GE cultivated maize. In 2015, 53.6 million ha of GE maize or 29% of 185 million ha (29%) of maize planted worldwide contributed to the global food supply. This was mainly in the United States with 33 million ha and Brazil, Argentina and Canada together with 17.4 million ha. In 2015, because of GMO productivity, 8 million ha (20 million acres) **were freed** for added food cropping thus contributing to citizens' food security. Recently, the 2018/2019 production of GE maize, was ~325 metric tons or ~29.6% of the worldwide maize production of ~1099 metric tons, and GE soybean production was 164 million metric tons or 44.6% of the global output of 367 million metric tons [31].

In 2019, 190 million ha (~469 million acres) in 29 countries (including 19 developing counties) were planted with GMO = GE crops, an increase of 5.7% of the total reported for 2015. A recent article lists countries and regions that do not allow GMO cropping or GMO food animals out of a concern for health safety. Some do allow non-food GMO cultivation and the importing of GMO grown corn and soybean for food animal feed [32].

Afterword

Increasingly warmer global temperatures during the decades preceding 2020 have resulted in changes to weather patterns, extreme weather events, and responses of lifeforms to warming of ecosystems that have damaged food production and hence food security in many regions. This problem and an approach to slowing and perhaps halting global warming and thus stabilize weather conditions and regularize food production are considered in the following chapter.

References

1. IPCC, 2019. Climate Change and Land. Chapter 5. Food Security. Online. www.ipcc.ch/srcc
2. UNFAO, 2019. The State of Food Security and Nutrition in the World 2019. Online. www.fao.org/state-of-food-security-nutrition
3. Met Office, Hadley Centre, 2015. Food security and climate change model spread. 23 p. Online. http://www.metoffice.gov.uk/food-insecurity-index
4. Met Office, Hadley Centre, 2015. Food security and climate change model spread. 23 p. Online. http://www.metoffice.gov.uk/food-insecurity-index/resources/…
5. Epstein, E., 1965. Mineral metabolism. In Plant Chemistry, Bonner, J. and Varner, J.E. (eds.), pp. 438-466, Academic Press, London.
6. Mosaic Company, 2020. Micronutrients. Online. www.cropnutrition.com/…/micronutrients
7. Hanif, A., Siddika, M.A., Mian, M.J.A., Hoque, T. and Ray, P.C., 2016. Effect of different micronutrients on growth and yield of rice. Int'l Jour. of Plant and Soil Sci., 12(6): 1-7.
8. United Nations Food and Agriculture Organization, 2019. The State of Food Security and Nutrition in the World 2019. United Nations, Rome. Online. www.fao.org/faostat/en…
9. Land Institute, 2019. Perennial crops: are hardware for agriculture. Online. www.landinstitute.org/our-work/perennial-crops
10. Kassam, A., Friedrich, T., Derpsch, R. and Kienzle, J., 2015. Overview of the worldwide spread of conservation agriculture. Field Actions Science Reports, Vol. 8. Online. www.rolf.derpsch.com/fileadmin/templates/main/downloads/…
11. Food and Agriculture Organization of the United Nations, 2009. How to Feed the World in 2050. Geneva, 35 p. Online. www.fao.org/…/How_to_Feed_the_World_in_2050.pdf
12. Blum, W.E.H., 2013. Soil and land resources for agricultural production: general rends and future scenarios - a worldwide perspective. Int'l Soil and Water Conservation Research, 1: 1-14.
13. Hunter, M.C., Smith, R.E., Schipanski, M.E., Atwood, L.W. and Mortensen, D.A., 2017. Agriculture in 2050: Recalibrating targets for sustainable intensification. Bioscience, 67: 386-391.
14. IPCC, 2019. Climate Change and Land. Chapter 1. Framing and Context. Executive Summary, 98 p. Online. www.ipcc.ch/sites/assets/uploads/2019/08/2b. Executive summary
15. Rare, T., Antonelli, M., + 7 co-authors, 2019. Transforming food systems under climate change: Local to global policy as a catalyst for change. CCFAS working paper 271, Netherlands, 35 p. + 3+ end notes. Online. www.transformingfoodsystems.com Scroll down and click on 'full report'
16. IPCC, 2018. Global Warming of 1.5° C. Special Report. Executive Summary. Online. www.ipcc.ch/sr15/download
17. FAO, IFAD, UNICEF, WFP and WHO, 2018. The State of Food Security and Nutrition in the World 2018. Building climate resilience for food security and nutrition. Rome, 181 p. (113 p text, 68 p annexes) Online. http://www.fao.org/3/19553EN/19553en.pdf
18. Pastor, A.V., Palazzo, A., Havlik, P., Biemans, H., Wade, M., Obersteiner, M., Kabat, P. and Ludwig, F., 2019. The global nexus of food-trade-water sustaining environmental flows by 2050. Nature Sustainability, 2: 499-507
19. Molotoks, A., Stehfest, E.,Doelman, J., Albanito, F., Fitton, N., Dawson, T.P. and Smith, P., 2018. Global projections future cropland expansion to 2050 and direct impacts on biodiversity and carbon storage. Global Change Biology, 24: 5895-5908.
20. Siegel, F.R., 2015. Countering 21st Century Social-Environmental Threats to Growing Populations. Springer, pp. 34-36 of 164 p.
21. Watson, A., Ghosh, S., Hickey, L.T., and 22 co-authors, 2018. Speed breeding is a powerful tool to accelerate crop research and breeding. Nature Plants, Vol. 4: 23-29.
22. Chen, S., 2019. Countdown starts for China's big mutant crop space mission in race for food security. South China Morning Post, 7 Nov.

References

23. Balastrini, R., Lumini, E., Borriello, R. and Bianciotto, V., 2015. Plant-Soil Biota Interactions. In Soil Microbiology, Ecology and Biochemistry, 4th Ed., E.A. Paul (Ed.), Academic Press, Chapter 11, pp. 311-338.
24. Riggio, J., Baillie, J.E.M., and 10 co-authors, 2020. Global human influence maps reveal clear opportunities in conserving Earth's remaining intact terrestrial ecosystems. Global Change Biology, 40 pages, National Geographic Society, Wiley Online Library. https://doi.org/10.1111/gcb.15109.
25. Duff, A. and Padilla, A., 2015. Latin America: agricultural perspectives. Series publication, Latin America after the commodity boom. Online. economics.rabobank.com/pub;ications/2015/…
26. Meyfroidt, P., 2018. Tradeoffs between environment and livelihoods: Bridging the global land use and food security discussions. Global Food Security, 16: 9-16.
27. UNFAO, 2018. Save Food: Global Initiative on Food Loss and Waste Reduction: Online. www.fao.org/save-food/resources/keyfinding/en
28. Kinley, R.D., de Nys, R., Vucko, M.J., Machado, L. and Tomkins, N.W., 2015. The red microalgae *Asparagopis taxiformis* is a potent natural antimethanogenic that reduces methane production during *in vitro* fermentation with rumen fluid. Animal production Science, 56: 282-289. https://doi.org/10.1071/AN15576
29. World Bank, 2019. Arable land (hectares per person). Online. http://data.worldbank.org/indicator/AG/LND/ARBL/HA/PC
30. US Department of Agriculture, 2019. Crop Acreage Data, Farm Service Agency. Online. www.fsa.usda.gov/…/crop-acreage-data/index
31. Pelligrino, E., Bedini, S., Nuti, M. and Ercoli, L., 2018. Impact of genetically engineered maize on agronomic, environmental and toxicological traits: a meta-analysis of 21 years of field data (1996–2016): Scientific Reports, Article 3113.
32. Genetic Literacy Project, 2020. Where are GMO crops and animals approved or banned? Online. geneticliteracyproject.org/gmo-faq/where-are-gmo…

Chapter 4
Impact of Global Warming/Climate Change on Food Security 2020

4.1 Introduction

From about the third quarter of the twentieth century to 2020, there have been increases of weather/climate related events (e.g., drought, heat waves, wildfires) that have disrupted food systems and hence food security for major segments of society. They have been especially injurious to economically disadvantaged populations. These events may be relatively short term or somewhat longer (e.g., a 2 week heat wave, a 7 day flood, a month of extreme cold, a month of wildfires, a 5 year drought). They also may be long-term continuously slow changing conditions. Included in this group are warming of the Earth, melting glaciers, observable and measurable creep of agro-climatic zones towards the poles with expansion of animal ranges (including insects), and invasive vegetation (weeds). There are migrations of terrestrial and marine animals from warming ecosystems that disrupt their life cycles to ecosystems with temperatures that favor their feeding and procreation. Together, global warming and resulting climate changes are the drivers of direct, triggered, and indirect weather events that lessen the world's human food carrying capacity and diminish ecosystems' natural resources capacities to meet societal needs [1, 2].

4.2 The Warming Status

The consensus from IPCC reports and other studies is that from pre-industrial times (1850–1900?), to 2015/2016, the Earth's mean temperature rose by about 1 °C. A 2017 study on the possibility of limiting global warming to the 2016 Paris Agreement target of 2 °C with the hope of limiting the warming to 1.5 °C posits that limiting post-2015 CO_2 emissions to a cumulative 200 gigatons carbon (GtC = 1 billion tons) through 2030 would limit warming to less than 0.6 °C in 2/3 of the computer simulations evaluated. This amount of warming added to the 1 °C rise since the pre-industrial era would keep warming within the 2 °C to 1.5 °C Paris Agreement target [3]. Another study used the little ice age (~1450) temperature as a pre-industrial era baseline that gave a higher rise of 1.2 °C to 2015. When the aforementioned rise of 0.6 °C is added to 1.2 °C rise and CO_2 emissions were kept at a cumulative 200 GtC from 2016 to 2030, there would be a warming to 1.8 °C, still within the Agreement target [4]. However, given the rate of emissions in 2017, 2018, and 2019, of 32.5, 37, and 33 GtC, respectively, with little future decreases foreseen, the cumulative 200 GtC will be exceeded by 2023/2024. In either case for the temperature selected as the preindustrial baseline, the warming has taken place and continues to increase. Graphs plotted for increases in **warming temperatures** and for **CO_2 concentrations** in the atmosphere since pre-industrial times to the present show a strong positive correlation thus corroborating the science based cause (CO_2 rise) and effect (continued warming) relation. Extrapolating the continued rise in global temperature from the datasets analyzed allows a prediction that the Earth's average surface temperature will reach and likely exceed 1.5 °C by 2050 if not sooner.

4.3 Extreme Weather Disasters Impact on Food Production

Climate forced extreme natural disasters have affected food production and consequent food security since farming began. These are extreme heat conditions over a relatively short period of time, droughts over long periods of time, floods following heavy rains or rain of long duration, periods of extreme cold and sometimes high winds. Heat and drought will dry vegetation and set conditions for wildfire disasters ignited by storm lightning strikes and ofttimes rapidly spread by high winds such as in 2019 and 2020 in South and Southeast Australia and Western United States.

In an investigation to determine global scale effects of extreme weather disasters (EWDs) on crop production, researchers evaluated 2184 floods, 497 droughts, 138 extreme heat, and 178 extreme cold disasters from 177 countries over a 43 year period from 1964 to 2007. Variations on the effects of EWDs depend on their severity (intensity), vulnerability (location), and exposure (duration). Researchers reported that drought and extreme heat **reduced crop production by 9–10% worldwide** with much of the lost crops at small farms in less developed and developing countries. They could not identify percentage production losses from floods and extreme cold. With more complete and improved data from 2000 to 2007, researchers estimated that **3.2% of the global cereal production was lost to drought and 3% lost to extreme heat**. From 2007 to 2013 there was an additional reduction in global production of 7% mainly from the loss and damage to crops in **developed countries** of 8–11%. The fact that developing countries have small farms and multi crops (some resistant to EWDs) vs. developed countries with extensive areas of mono crops that can be wiped out by an EWD may explain the developed countries higher percent loss. Overall, the estimated **losses from drought** during the 1964–2007 time span evaluated was 1820 million tonnes, an amount equal to the total global 2013 wheat and maize production, while 1190 million tonnes or an amount equal to the 2013 global maize production was **lost to extreme heat events** [5]. Adaptations to lessen the disruption of food production from EWDs (e.g., improved drainage and flood control, heat and/or drought resistant crops) can support improved food security for at risk populations worldwide. Examples of EWDs that hurt food production are described in the following paragraphs.

4.3.1 Examples of Extreme Weather Disasters and Their Effects on Food Supplies in Recent Decades

Drought has been instrumental in the deaths of thousands of heads of cattle and other livestock in the past and in recent years. For example, the 7 year Federation Drought in Australia, 1895–1902 destroyed 40% of the sheep and cattle stock and essentially destroyed the Australian wheat crop. Fast forward to 2014–2019 in Queensland where 5 years of extreme drought stressed herds to the degree that they could not withstand the monsoon rains that flooded Queensland with 3 years worth of average rain in a week killing more than 600,000 cattle and 48,000 sheep [6].

In addition to killing livestock, recurring heat waves and drought in South and Southeast Australia set conditions for wind-driven seasonal fires that burn down forests and their animal habitats, buildings, scorch grazing land, and burn feed crops and stored hay that sustain livestock. In some cases, it is feasible to move livestock to locations where they can find water and feed, otherwise they perish. It should be noted that another Australian disaster followed that described in the previous paragraph. Bushfires at the end of 2019 that extended into 2020 scorched 11.4 million acres killing 1.7 million sheep and 450,000 cattle, and burning 5900 buildings including 2800 homes. This affected the food system (meat, fruit, dairy production and distribution) and national exports (wool, wine, cotton, and honey). In forests and fields, the fires killed as many as one million wild animals [7].

A severe drought in Argentina during 2008/2009 significantly lowered yields of the feed grains corn and soy for both export and domestic use. The lack of sufficient feed and water led to the slaughter of thousands of cattle including female cows thus leading to smaller herds in following years (Table 4.1). Less need for transport, storage, and other parts of the food system resulted in loss of income for employees and taxes for the state as well as loss of export fees that had been loaded on farmers by the 2008/2009 Argentine government [8].

Another severe drought in Argentina during the latter months of 2017 into 2018 contributed to the deaths of as many as a million head of cattle. As in 2007/2008, soy and corn yields were greatly decreased. In addition, the drought caused significant decreases in chicken and pork production.

Table 4.1 Cows/calves evolution 2007–2010 illustrating effects of a 2008 severe drought in Argentina as head counts, in millions of heads, during and post-drought dropped significantly [8]

	2007	2008	2009	2010
Cows	24.4	23.9	22.2	20.5
Calves	15.4	15.4	14.1	12.6

During July, 2019, heat waves that punished Europe caused intense drought conditions in France that reduced the output of one of the continent's largest grain producers. Heavy rains and consequent flooding soaked fields during the 2019 spring planting season in large areas of the midwest United States farm belt and caused delays in the sowing of soybeans and corn.

African nations are especially vulnerable to EWDs that hurt crops because the continent depends to a great extent on subsistent farming with an estimated 70% of the people that grow a good amount of their own food. For example, in the Horn of Africa (Eritrea, Ethiopia, Djibouti, and Somalia), a drought in 2019 followed by heavy rainfall decreased crop yields and livestock herds to the degree that 22 million people had severe food insecurity. The rains and flooding in the region disrupted the transportation infrastructure thus halting delivery of food that was available. Add to this that some nations were not financially able to import food for their populations where crops failed. Away from the Horn of Africa, drought has caused severe food insecurity because of reduced and failed food crops in Zimbabwe and Zambia and with western Kenya reporting a 25% food crop loss [9].

4.3.1.1 Special Insect Problem: Locust Swarms

In 2020, as a result of extreme weather with much rain in warm areas, enhanced insect breeding conditions brought a special threat to crops that tens of millions of people in Africa depend on for food security already being undermined as described in the past paragraph: plagues of locusts swarms that consumed crops in parts of Kenya, Somalia, and Ethiopia. Kenya had not experienced such a locust invasion for 70 years. A 1 km^2 swarm can have 40–80 million locusts that in 1 day can consume the same amount of food as 35,000 people. Multiple swarms can disrupt food security, often in regions with high population growth. Locust infestations also reduced crops in Uganda, across SW Asia (India, Pakistan), and the Middle East (Yemen, Saudi Arabia, Iran). Locusts watchers, some recently using drones, work to identify where breeding is taking place and use chemical pesticides sprayed from trucks or planes to try to control their numbers but the effects of these pesticides on biodiversity and humans is in question. A more directed control is by using natural means by making killer powders from bacteria, fungi, or viruses, mixing them with oil and spraying targeted breeding grounds when they are located [10].

4.4 Observed and Measured Ongoing Effects of Global Warming on Ecosystems and Food Production

Global warming and the climate change it causes has multiple interrelated effects on crop production and food carrying capacity to sustain the Earth's growing populations. A positive impact may be that growing seasons can start earlier and last longer in changing agronomic zones. A longer growing season may allow for double planting of the same crop or a different one. This would add to food production and food carrying capacity. A negative result is that there are changes in agronomic zones as warming moves poleward (in both hemispheres) and may necessitate changes in which crops are planted. Negative effects are that a warming environment (poleward) can result in weed migration and enhance their growth as they rob soil of nutrients. The warming also favors newer and higher insect/pest populations that can damage crops or bring disease to them. These require control if food production is to be maintained and increased. Some animals and plants migrate from warming habitats to which they cannot adapt and sustain their life cycles by moving to habitats with cooler temperatures where they can thrive. Other animals and plants react to warming by expanding their ranges and the breadth of habitats that support them. Such changes affect animal/plant abundances and seasonal activities that can disrupt or support procreation and food chains on land and in oceans/seas such as for insects, deer, rodents, weeds, polar bear, fish, shellfish, and sea algae. This in turn can disrupt food production and security.

Food production and security have also been threatened from global warming driven changes in rain patterns experienced during past decades (location, timing, and amount of rainfall) with wet areas getting wetter and dry areas receiving less precipitation. A result is that with shifting of climate zones poleward, there is **more rain in the middle and high latitudes that leads to an increase in yields of some crops (e.g., maize, wheat, sugar beets)**. However, **in low latitude regions (tropics and sub-tropics) with less precipitation, yields of some crops decline (e.g., maize, wheat)** [10, 11].

As will be discussed in a later chapter, computer models forecast that future land degradation (e.g., desertification, erosion) will be at moderate to high risk with global warming of 1.5–2 °C. This degree of warming on land and in water bodies together with more intense, more frequent and longer lasting extreme weather events are likely to disrupt food systems and lessen food security for the Earth's growing population. There is a possibility of greater warming yet to 2.5 °C or 3 °C or higher above the pre-industrial baseline if CO_2 emissions to the atmosphere are not significantly reduced as the second half of the century progresses. This raises 'a red flag' about future food production capacity and security (discussed Chap. 8).

4.5 Counter the Effects of Weeds, Insects/Pests, Disease

Warmer temperatures have created conditions for increased growth of weeds and the proliferation of pests affect plantings. Climate changes in growing zones will likely require the use of herbicides and pesticides to counter invasive weeds and insects/pests. Pests are currently responsible for 25–40% of all crop loss (e.g., the yield, nutrition value, and outward appearance [salability]). The warming can change the migration times of birds (predators) that otherwise keep insect populations (preys) down and reduce the damage they may do or disease they may impart to a crop [12].

Even with crop loss and with waste of foods, higher crop yields and expanded cultivated fields have produced enough food to sustain the 2020 ~7.8 billion human population. However, as noted earlier in the text, the 'enough' food has not reached the 800 million to 1 billion people on Earth that remain undernourished/malnourished because of poverty, infrastructure failures, political decisions, and conflicts/wars. The question that will be discussed Chap. 8 is whether production will increase enough in the future on a warming Earth to feed an additional 2.3 billion people in 2050, or an unlikely but potential population of 11+ billion in 2100. With decadal global warming proceeding at 0.15 °C, this would require food production increases from climatic zones amenable to improved crop yields and from controlled farmland expansions.

In lieu of herbicides and pesticides, some farmers opt to sow genetically modified (GMO) seeds that protect crops against pests, herbicides used against weeds, and/or resist damage from earlier cited **extreme climatic conditions**: extended droughts and heat waves that besides being more intense, may be more frequent and long lasting. This is predicated on availability of GMO seeds with desired properties and on national decisions of whether or not to allow their use. Where a warming climate could expose crops to a pest borne disease, chemical sprays can be used, or if available, a farmer can sow seeds altered by GMO to eliminate or mitigate the effect of a **specific disease**.

4.6 Warmer Temperature Effects on Crop Physiological Growth

A warming agricultural zone can cause changes in the make up of living organisms (e.g., their cells, tissues, organs, and/or a total body system) and how they function with respect to changes in physical and chemical factors (physiological effects) that can harm crop development. For example, from the physiological point of view, most plants function normally at temperatures up to 40 °C, but not fruits. Fruits are known to mature normally at temperature as high as 35 °C. However, higher temperatures, especially in tropical and semi-tropical regions, can block the ripening process and can cause faster tree maturation (earlier budding) of fruit that will be slow to set. This can result in loss and waste in crop production from poorer yields and quality (nutrition and appearance). Reduction of these effects on fruits and other food crops by improving resistance to higher temperatures via hybridization or genetic modification can maintain crop sustainability and decrease the need for cropland expansion to maintain production. On the positive side, a warming that is earlier and does not exceed 35 °C may result in faster tree maturation of fruits which results in earlier harvest date and sooner export to markets. Chapter 8 will explore the effect of projected warming to 1.5 °C and 2 °C and higher on the grain crops maize, wheat, rice, and soybean (Tables 8.1 and 8.2).

4.7 Survival Plan: Reduce CO_2 Emissions

The obvious answer to slow the rate of global warming and the climate changes it engenders is by adapting to changing ecosystems while working to mitigate the human activities that drive the warming. An achievable aim is to **slow the Earth's rate of temperature rise** (now 0.15 °C/decade) but this would require **total international cooperation** and investments in technological advances, conditions that do not exist in 2020. If this cooperation aim is met in the foreseeable ('near') future, a followup, also achievable with adherence by all nations, costs not withstanding, is to **stabilize the Earth's temperature**, a goal of the Paris Agreement. The **Gordian knots** to reaching the goal of stabilization are the drive to industrial economic development by many countries that itself functions to force global warming, and global population growth with increased demands for manufactured products, even as the needs for sustainable sources for water and food are not being met for all citizens in many less developed and developing nations. With stabilization of the Earth's temperature at a 'livable' level, the global citizenry would have to live with a new 'normality' in much the same way as the 2020/2021 global population is living a new normality with the COVID-19 pandemic. An ultimate goal would be to reverse warming, an achievable goal for our present global society only with an 'all in' commitment following 'what can be done' protocols.

What can be done to reduce the mass of CO_2 emitted from fossil fuels and other sources worldwide in coming decades? Can the CO_2 be captured at industrial sources and used or safely disposed of and can it be reduced at agricultural sources, or can the CO_2 be sucked out of the atmosphere to be used or stored/disposed of safely to slow and finally stop global warming? The answers are yes, it can be captured, it can be used, and it can be stored in deep underground rock formations. Chemical scrubbers in industrial operation, if used and maintained, can capture 80–90% of CO_2 being produced. Of the captured CO_2, most has been used as a medium for secondary recovery at oil wells (EOR, enhanced oil recovery), some to make salable products, and some stored underground in porous, permeable geologic formations confined by impermeable rock that prevent leakage to the surface. With respect to the latter, a recent paper reported that based on 22 years of CO_2-EOR operations in a Northern Michigan Basin from 1996 to 2017, large CO_2 storage volumes in limestones at depths of 1500–2000 ft resulted in net **negative** emissions of greenhouse gas. This is due in part because the CO_2 used is separated from a natural gas facility and not vented to the atmosphere [13]. Vehicle CO_2 emissions will be greatly lessened as more EVs, HEVs and PHEVs are sold as battery prices come down, ranges are increased, and charging stations (renewables powered) are set up to support distance driving. Quick charge batteries (15 min vs. 8 h for full charge) for cars and trucks that will abet long trips are being pilot studied with field tests by StoreDot. This can significantly reduce CO_2 emissions. The question remains as to whether (or if) the CO_2 emissions capture programs now functioning in part in many countries can be activated globally soon enough before reaching a tipping point when warming leads to a "heat pandemic" that is too much for all populations to withstand. If CO_2 emissions into the atmosphere are not reduced so that global temperate is stabilized by 2050–2060, many people are likely to die from a government inability to meet the demand for water and food, and cooling to prevent death from excessive heat exposure. This could happen in those regions without the agro-technology and/or water resources that can develop enough food production capacity (and delivery network) to sustain growing global populations and those without the financial capacity be able to cool people…in less developed and developing countries…especially in Africa, Asia, and somewhat in South America.

4.7.1 Removal of CO_2 from the Atmosphere?

There are several international companies in Canada, Norway, United States, Netherlands, Poland, Switzerland that have research and development programs and planned or pilot projects to suck air from the atmosphere with systems adaptable to use renewable electricity (solar, wind, hydropower, geothermal, nuclear?) to separate carbon from the inflow of CO_2. One company proposes to react carbon with hydrogen (e.g., separated from water vapor, from ammonia) to manufacture low-carbon jet fuel [14]. Royal Dutch Shell went from proposal to practice and produced kerosene from CO_2 and water (source of hydrogen, H_2) to serve in a jet fuel mix to reduce CO_2 emissions from aircraft to help slow global warming. Five hundred liters (132 gallons) of the kerosene was included in the fuel for a January, 2021 KLM flight from Amsterdam to Madrid and return. Air-captured CO_2 is being commercialized on a small scale in Switzerland to make carbonated sodas and is available for use in greenhouses to stimulate growth in plants/vegetables/fruits being cultivated for sale. Other products made at industrial sources with captured CO_2 are compounds such as sodium and magnesium carbonate and sodium bicarbonate (baking soda). An algal biofuel program is researching the used of sodium carbonate in its process because it speeds up algae maturation and enhances the formation of lipids [15]. The reality of the impact of extracting CO_2 from air on global warming, given 2020 capabilities, is further explored in Sect. 8.7.1.

The mass of CO_2 in the atmosphere is ~3×10^{12} tonnes (tera, T, trillion), with an estimated CO_2 2019 emission of $36.8 \times 10^9 \pm 1.8$ tonnes (giga, G) [16]. To have a chance at meaningful reduction in the atmospheric CO_2 content in addition to what could be achieved by reductions from sources such as industrial and vehicular sources, and direct air capture cited previously would also require "sink" preservation and growth. This requires mitigation of forest/vegetation loss that damages CO_2 sinks. It means careful planning that outlaws clear cutting, mandates the planting of seedlings, and support of biodiversity. New growth can rise by natural seeding from protected stands of trees and from animals that ingest tree fruits and leave seed bearing droppings on soils. Global reduction in CO_2 released from agriculture (cropping protocols and animal husbandry) would add to the effort to lessen the CO_2 emissions to the atmosphere. At present, agricultural efforts have reduced CO_2 and CH_4 (flatulent methane) emissions. In some major forested areas (e.g., Indonesia in 2019 and 2020) management is now active in preserving their "sinks" but better forest management in Africa and in Brazil has to stop rogue timbering and government supported forest clearing for economic reasons. As reiterated in other chapters, **the future is now** for instituting actions that that can slow and ultimately stop the warming and its negative impacts on societal needs lest we pass environmental tipping points that will bring nations and billions of people to a pandemic-like condition for which there is no vaccine.

Afterword

The warming of the Earth and consequent climate changes have reduced the amount (yields), quality (nutrition), and location of food production in many parts of the world. Agro-technology has advanced improvements so as to counter those factors that lead to loss of food production while supporting research into others that can sustain and increase food production. The future of global food security later in this twenty-first century (2050, 2100) as populations grow rapidly in Africa and Asia is explored in Chap. 8. Beyond food security and access to safe water as basic needs to sustain life for the Earth's population, sanitation, the subject of the next chapter, is a requisite that is fundamental to sustaining public health and hence citizens' contributions to societal development.

References

1. EPA, 2017. Climate impacts on agriculture and food supply: crops livestock, fisheries. Online. 19january2017snapshot.epa.gov/climaye-impacts-on-agriculture…
2. Simmons, D., 2019. A brief guide to impacts of climate change on food production. Online. www.yaleclimateconnections.org/2019/09/a-brief-guide-to-the-impacts

3. Millar, R.J., Fuglestvedt, J.S. and 8 co-authors, 2017. Emission budgets and pathways consistent with limiting warming to 1.5°C. Nature Geoscience, 10: 741-747. Online. http://eprints.whiterose.ac.uk/120597/3/millaretal_natgeo
4. Schurer, A.P., Mann, M.E., Hawkins, E., Tett, F.S.B. and Hegeri, G.C., 2017. Importance of the pre-industrial baseline for likelihood of exceeding Paris goals. Nature Climate Change, 7: 563-567. Online. http://centaur.reading.ac.uk/71780/1/schurer_etal_2017_NCC…
5. Lesk, C., Rowhani, P. and Ramankutty, N., 2016. Influence of extreme weather disasters on global natural cereal crop production. Nature, 529: 84-87.
6. Major, T., 2019. Australia counts the cost of wildfire cattle, infrastructure losses following Queensland floods could near $2b. ABC Rural News, April, 15.
7. Business Times, 2020. Australia counts the cost of wildfire damage to crops and livestock. January 13. Online. www.busiesstimes.com.sg/consumer/australia…
8. Gevara, J.C. and Grunwald, E.G. 2012. Status of Beef Cattle Production in Argentina Over the Last Decade and Its Prospects. In Livestock Production, INTECH OPEN, Chapter 6, pp. 117-134. Online. https://doi.org/10.5772/50971
9. Childs, J.W., 2019. More than 50 million people face hunger crisis due to the African drought. Online. weather.com/news/news/2019-11-08-africa-drought…
10. IPCC, 2019. Climate Change and Land. Chapter 5. Food Security. Online. www.ipcc.ch/srcc
11. Willett, W. and 36 co-authors, 2019. Food in the Anthropogene: the EAT-Lancet Commission. Healthy diets from sustainable food systems. Lancet, 393: 447-492
12. UNFAO, 2020. Biocides for locust control. Online. www.fao.org/fao-stories/article/en/c/1267098…
13. Sminchak, J.R., Mawalkar, S. and Gupta, N., 2020. Large CO2 storage volumes result in net negative emissions for greenhouse gas life cycle analysis based on records from 22 years of recovery operations. Energy and Fuel, 34: 3566 - 3577.
14. Johnson, E., 2019, Reclaiming fuel from air. Chemistry World. Online. www.chemistryworld.com/news/reclaiming-fuel-from-air
15. Wesoff, E., 2010. Skytronic cleans up coal: update. Online. www.greentechmedia.com/articles/read/Skytronic
16. Muti-Authors, 2019. Global Carbon Budget 2019. Earth Syst. Sci. Data, 11:1783-1838. Online. https://doi.org/10.5194/essd-11-1783-2019

Chapter 5
Sanitation, Waste Generation/Capture/Disposal Status 2020

5.1 Introduction

The Earth's carrying capacity of human, animal, and plant populations is the maximum population size of a species that an area (environment, ecosystem) can support without reducing its ability to support the same species in the future (indefinitely). This is true in terms of clean air, safe water, and nourishing food for which cleanliness (free of harmful pollutants) supports disease prevention and populations' good health. Wastes from these populations, especially from humans can be a problem.

Humans and other lifeforms can not survive in their own wastes and more so when added to by wastes originated, for example, from domestic/commercial, manufacturing and industry, agriculture, and energy/transportation sectors. We may ask then "does the Earth's have limits to its capacity to absorb wastes as solids, liquids, and gases, organic and inorganic, from these and other sources without threatening global carrying capacity to sustain a reasonable quality of life for human populations and other lifeforms in their ecosystems?" Basic to the carrying capacity is access to clean water, proper toilets or latrines, and hand washing but not too far behind these sanitation basics are those that give health security from human activities generated wastes cited above.

5.2 Open Defecation: A Problem and a Solution

In 2017, there were more than 2 billion people of the ~7.6 billion worldwide that did not have access to basic sanitation facilities (toilets or latrines). Of these, an estimated ~892 million (812 million in rural areas) practiced open defecation (in street gutters, in the bush, or in open waterways). Another 856 million used unimproved sanitation facilities such as pit latrines or latrines without a slab. Most open defecation takes place because of lesser stages of national development, much poverty, and in some cases, conflict violence and weak central governments.

Of the main populations that practice open defecation, 558 million are in low and lower middle income countries in Central and Southern Asia, and 220 million in Sub-Saharan Africa [1]. There has been progress in eliminating this source of disease. From 2000 to 2015, the Asian rate of open defecation dropped from 53% to 30% and in Sub-Sahara Africa it dropped from 32% to 23%. However, as a result of the high birth rate in Sub-Saharan countries and increased populations, the actual number rose from 204 million practicing open defecation to 220 million.

Feces release pathogens from open defecation in the bush, in agricultural fields where they may access soils with possible transfer to food crops, seep through soils and underlying rock into aquifers, and drain into open canals and waterways. The pathogens may also be moved by rain water or floods from land into waterways infecting them with bacteria that may be transferred to humans via drinking, cooking, washing, or crops irrigated with contaminated waters. This lack of basic hygiene causes transmission of diseases. For example, poor sanitation is considered to be one the main causes of 432,000 deaths annually, 297,000 of which are children under 5 years old. As noted previously, open defecation takes place most often in rural areas but can be a major problem as well in mega-cities that house tens of millions of people and in other highly populated cities where many economically disadvantaged citizens, especially those living in the peripheries are without access to improved sanitation conditions/facilities: clean water, flush toilets tied to sewer systems, septic tanks, pit latrines (preferably ventilated and on slabs), or hand washing stations. In our world of increasingly populated urban centers, sewer systems are necessary to receive human excreta and transport it to treatment plants. Here, the first stage screens out the coarse solid matter and then the small size solids. Next there is a bioremediation with aerobic or anaerobic bacteria depending on the characteristics of fine wastes and water chemistry and the output is then passed successively through coarse, medium, and fine size sand to remove odor, flavor, and color. The cleansed water is lastly disinfected with chlorine or hydrogen peroxide to eradicate any bacteria or viruses that survived the previous treatment and either release a clean effluent into waterways or put it back into clean water use distribution network.

5.2.1 Reducing Open Defecation with Education

Education of those practicing open defecation is the way to stop the practice in favor of the use of honey buckets that are collected daily for safe disposal and the use of latrines and flush toilets as they may become available. This begins with a government commitment to support projects based on community education. Education has to be presented to communities in the plain daily language people relate to for clear understanding of the importance of changing this often cultural habit in order to eliminate sicknesses (mainly diarrheal) from feces pathogens contaminating water used for drinking, cooking, and personal hygiene. In India, diarrhea kills over 117,000 children under 5 years old annually with an additional 60,000 children in the other countries of South Asia [2, 3]. Diarrhea and other sicknesses from fecal contaminated water keeps others ailing in their homes thereby losing school days for children and workdays for adults, thus harming local and national economies. The use of improved toilet facilities together with simple hygiene such as hand washing can basically eliminate human waste related problems in a community. Educational programs have been successful over the years in Bangladesh and Viet Nam when they included engineers who explained and showed communities how to build sanitary latrines, where to build them so as not to contaminate an aquifer or surface water supply, and how to maintain or repair them when necessary.

In Bangladesh, the government invested 25% of its development funds in such an effort and reduced the incidence of open defecation from 43% of the population in 2003 to ~1% in 2015 and basically to 0% shortly thereafter. This reduction of open defecation and sick days lost to diarrhea, dysentery, parasitic worms and other illnesses, plus a decline in population growth have contributed to a rise in the Bangladesh GDP growth rate from 4.7% in 2003 to 6.6% in 2015 and 7.9% in 2019 with a corresponding rise in family income and number of families moving out of poverty. In Viet Nam, from 2000 to 2017, the per cent of people practicing open defecation dropped from 21.8% to 3.9% in rural areas and from 5.1% to 1.3% in urban centers for a national average of 2.99%. The country aims to be open defecation free by 2025 if not sooner.

5.3 Food Animal Wastes

As important as is good sanitation for the health of human populations, sanitation applied to food animal wastes from high volume/high density commercial feedlots is similarly important to support the Earth's human carrying capacity. This is from the aspect of food security and from the possible impact of animal waste on humans' health and ecosystems that sustain human needs. The problem of animal waste becomes evident if we consider that the daily total solids manure from one head of young feeder cattle is 15× that of a human [4]. Thus, a feedlot with 100,000 bovines in Texas, USA can generate the same daily fecal waste as 1.5 million people. In the state of Texas, USA there are 2.7 million head of cattle in feed lots so that the daily mass of total solid fecal matter could equal to that generated by a human population of 40.5 million or almost twice the 2019 Texas population. There are 32 feedlots in the state of Kansas, USA with 50,000 or more cattle so that the total solid fecal waste generated daily would be that of a mega-city with 24 million people, or more than 8× the state population.

5.3.1 Magnitude of the Global Food Animal Waste Problem

There were 1.468 billion cattle worldwide in 2019. Of these, more than 48% were in Brazil (212 million), India (189 million), China (113 million), the United States (89 million), Ethiopia (54 million) and Argentina (51 million). About 18% of these or 265 million head are milk cows. Also, there were 978 million pigs worldwide in 2018. Of these more than 86% are in China (447 million), the European Union (150.3 million) and the United States (74.5 million) [5]. As a result of the African swine flu epidemic that hit Asian pig farms in 2018 and continued into 2019, millions of hogs have been culled (Table 5.1) and as of September, 2019 continue to be culled. Added to fecal waste from these sources, there is fecal waste worldwide from sheep flocks numbering 1.21 billion animals and goats herds numbering 1.04 billion animals. These FAO numbers are not always in agreement with estimates from other sources but are of the correct orders of magnitude. Clearly, the mass of food animal fecal waste to rid of coliform bacteria before being used or disposed of poses a major problem for the 2020 world population that could be more of a threat to increased populations in 2050 and possibly later in the century, especially in less developed and developing countries that are exporters of the raw (frozen) animal meats.

In general, the animal populations cited above are in locations away from large human populations. Nonetheless, disposal of animal wastes has to include **cleansing them** of coliform bacteria and residues of growth hormones (anabolic steroids) and antibiotics they were fed or injected with in many high volume, high density feedlots or chicken farm operations. This will reduce the possibility of water and soil contamination that could otherwise be a threat to clean water and food security. Disposal by spreading the feces on open land stopped when the volume became too great for the environment to absorb. A preferred method now being used depends on manure fermentation and temperature increase in manure stacked as tiered masses close to feedlots. The internal temperatures developed, >131 °F but <160 °F, kills contained pathogens but does not harm beneficial microbes. A problem here is that the edges of the stacked manure do not rid it of the pathogens and have to be reprocessed [7, 8].

An improved fermentation method to rid manure of its pathogens is to compost it in large rotatable drums that hold 1000s of pounds of manure and that are turned once a day, mixing and aerating the mass evenly allowing it to heat to >131 °F but

Table 5.1 African swine flu infected pigs in Asia culled by September, 2019 [6]

Country	Culled numbers
Viet Nam	4.7 million
China	1.2 million
Laos	25,000
South Korea	15,000
Hong Kong	10,700
Philippines	7950
Mongolia	3115
Cambodia	2400

<160 °F eliminating >90% of the pathogens while reducing the manure mass weight by >60% to nutrient rich matter. This is dried, bagged, and sold as fertilizer that also enhances moisture retention in a soil. The dried manure can be pressed into briquettes that can be burned to generate energy.

A problem with the fermentation approach is that methane, a potent greenhouse gas, is released to the atmosphere thus supporting global warming. To get around this, a method followed in some rural areas of Asia and Africa is to tap large drums of composting manure for methane gas as an energy (cooking) source. The UK is experimenting with this technique on a larger scale and is now feeding enough 'bio-methane' into the UK gas grid to support natural gas use in ten homes. However, the UK is unlikely to expand this application because it emits CO_2 to the atmosphere.

5.3.2 Using Slaughter House Wastes

Another source of major food animal waste matter that can impact the earth's human carrying capacity with respect to sanitation is from slaughter houses. This includes blood, wash water, heads, hooves, innards, skin, bones, and feathers. If isolated, treated, and used, or disposed of securely, these wastes will not contaminate surface or aquifer clean water sources or agricultural areas. The enormity of this source of waste produced is understood from the number of land food animals killed for food globally in a year. In 2016, these were 66 billion chickens, 1.5 billion pigs, 550 million sheep, 450 million goats, and 300 million cows [9].

To cope with this food animal waste problem, wastes can be treated to kill pathogens and processed to make salable byproducts that add to slaughter house owners incomes. The wastes are processed differently for specific products. These include feed for pets and animals or aquaculture farms (protein-rich meat meal, carcass meal), fertilizer for agriculture (bone meal), and tallow (fat). In addition, the anaerobic decomposition of viscera in compost systems as cited previously for manure, can generate methane gas that can be tapped as an energy source and used on site [10, 11]. In addition to composting, some solid wastes can be incinerated as necessary, if for economic or other reasons, other methods can not be used. Not all meat/dairy producing operations have the options to treat and process all their wastes, thus putting environments at risk if unused wastes are disposed of at land fills that are not of the sanitary type that protects water resources and proximate land areas from pollution that dumps/landfills can release during rains and runoff. The EU dealt with solid waste disposal by passing legislation in 2009 that banned the land disposal of all animal wastes except for manure and undigested feed. As populations grow from ~7.8 billion in 2020 to the ~9.9 billion estimated for 2050, and more people move into the middle class in developing and less developed countries, the demand for animal-sourced products will grow and output from slaughter houses will increase along with their mass of wastes generated. We can ask if the Earth can develop the capacity to absorb an added waste output if not all producers treat and process it? Dumping in the oceans is not a viable long-term option. Unless investments in education and in treatment/processing facilities is made that would yield profits for an investor, the waste problem will persist and grow especially for lower and lower middle income countries.

5.4 Pollution (Waste) Endangers Earth's Human Carrying Capacity

Pollution can be defined as physical, chemical, and/or biological contamination of an environment. This may be from **human originated waste** (e.g., excreta or industrial/manufacturing released toxins) **and naturally released toxins** (e.g., from an unexploited ore deposit) that threatens the health and welfare of humans and other lifeforms habituating there.

Pollutants can limit or reduce the Earth's human carrying capacity in local, or rarely, regional areas by contaminating a clean water supply, by contaminating food that takes up toxins from soils, and by diminishing useable food production (security) through irrigation of food crops with polluted water. In unusual happenings, pollutants are inadvertently released from natural sources by human activity and cause sickness. This happened in southwestern India and southeastern Bangladesh where millions of people were exposed to arsenic (As) poisoning from well waters. The As was present in the mineral pyrite (iron sulfide $Fe[As]S_2$), an impurity in the region's aquifer rock. The As was released when seasonal over pumping of aquifers lowered the water table and exposed the mineral to aeration (oxygen) that caused pyrite decomposition and an As release into aquifer waters used for irrigation as well as for drinking and cooking. Over time, thousands of people in the two regions bioaccumulated the As toxin in their systems from their water and food and suffered serious sicknesses [12–14]. This was a signal event because it made the WHO and most countries readjust the permissible levels of As allowed in drinking water to 0.01 mg/l (=10 µg/l or ppm).

Potentially toxic metals and outdoor and indoor air pollutants have been released into environments from various industrial point sources worldwide with coal-fired power plants, manufacturing and recycling facilities, smelters, and mining operations being among the worst offenders. Legislation has been passed in most countries that require the installation and use of emission and effluent control and capture of metals from industrial wastes for later treatment, reuse/recycling, or secure disposal. This is complemented by control, capture, and secure disposal of fine size particulates (<2.5 µm and <10 µm) that are known to harm lung functions if released to the air. The particulates can also react with industrial and vehicular emissions to create smog that can sicken and kill those that are exposed to it during an event especially if they already suffer from respiratory illnesses or have cardiac deficiencies. An existing failing in some countries is the non-enforcement or token enforcement of laws requiring (efficient) use of control/capture equipment and its maintenance because of a government's drive for economic development or

5.4 Pollution (Waste) Endangers Earth's Human Carrying Capacity

corruption of officials. Nonetheless, as governments and investors are shown that healthy work forces and families improve productivity and hence sustainable or increased profits for companies and tax income, the implementation of controls to diminish global pollution from metals and particulates has increased. This bodes well for the future even as demand for services and goods expands with world population growth during the coming decades (to 2050 and perhaps beyond) [15].

5.4.1 Can We Sanitize or Must We Depopulate Areas Evacuated During/After a Major Radiation Release?

Radiation released to the living environment from damaged nuclear power plants in Chernobyl, USSR (now independent Ukraine), and in Fukushima, Japan, has diminished the Earth's human carrying capacities of important areas that previously supported people's occupation, use of living space, and agricultural productivity. The 1986 Chernobyl disaster caused the evacuation of 350,000 people from a designated exclusion zone that was initially set as 2600 km^2 (1000 mi^2) but was subsequently extended to 4143 km^2 (1600 mi^2). The exclusion zone will be uninhabitable for 100–300 years depending on the concentrations and half-lives of the radioactive isotopes Cs^{137} and Sr^{90}. Much of the exclusion area was productive farmland that no longer contributes to food carrying capacity in the region [16]. As the result of the 2011 nuclear plant accident at Fukushima, the initial exclusion zone living space was 807 km^2 (311.5 mi^2) that has been reduced to 371 km^2 (143 mi^2). Initially 150,000 people were evacuated but a reduction of the exclusion zone left half that number displaced [16]. Worldwide there are 449 nuclear power facilities with 58 under construction and another 154 in the planning stage. In 2018, nuclear power supplied more than 10% of the world's electricity and 18% in the 36 OECD countries (with France most invested by generating 74.8% of it's electricity from nuclear power) [17]. Nuclear power does not emit CO_2 emissions that fuel global warming and will continue to contribute to the Earth's electricity generating capacity.

Radioactive wastes from the 449 nuclear power facilities worldwide are stored in water cooled containment pools at the same facilities where they have been generated. In the 70 years of nuclear power generation few disposal sites globally have been found geologically suitable to lock the wastes up for the tens to hundreds of thousands of years necessary for deadly radioisotopes to decay to levels that do not present a threat to human populations. These sites are in development. Add to this the wastes generated by the manufacture of nuclear weapons and naval vessels powered by nuclear propulsion by at least nine nations and the radioactive waste problem becomes more daunting. The secure storage/disposal of radioactive waste should be a major priority for nations/regions housing temporary storage sites.

5.4.2 CO_2, a Pollutant?

On the global scale, CO_2 naturally present in the atmosphere of the pre-industrial world at a concentration of 285 parts per million (0.0285%) has since increased by April 2019 to 413.33 ppm and then rose to 416.21 ppm in April 2020.

During the first 6 months of 2020, less CO_2 was emitted during the same period in 2019 as a result of the pandemic. The decrease of 1.55 billion tonnes reduced annual emissions such as the 37 billion tonnes emitted during 2019 by a few percent but cannot be expected to continue after the pandemic is 'under control' and pre-pandemic activities resume 'normal' functioning. The drop in emissions is attributed in grand part to transport emission that decreased by 40% (people teleworking from home), power emission decreased by 22%, and industrial emission decreased by 17% [18].

As the principal component of the greenhouse gases in the atmosphere that abet global warming and force changes in the Earth's climate, **CO_2 is an emitted waste product** from multiple sources. Can it then be considered a contaminant? This text's answer is yes because global warming and resulting climate changes threaten the health and well being of humans and other lifeforms from one or a combination of events. These we know are rising sea level and extreme weather events such as drought, heat waves, tropical and higher energy storms and their accompanying storm surges that are more frequent, more intense, and of longer duration. All such events lessen the Earth's carrying capacity for essential to life commodities and living space. As previously cited, animal and plant migration and agricultural productivity are other results of climate change. Global warming and climate changes also leave populations with an increased vulnerability to infectious disease carriers, and non-communicable diseases. This CO_2 'contaminant' problem needs attention and action focussed on solution if populations in future decades of the century are to have a reasonable quality of life whether in less developed, developing, or developed countries. A solution can be set in four tracks: (1) reduce the industrial sector emissions of CO_2 into the atmosphere—doable with existing technologies, (2) decrease the loss of great expanses of forest/vegetation CO_2 sinks while reestablishing illegally exploited and damaged former sink areas—doable, (3) increase the use of electric and hybrid electric vehicles—being done, and (4) suck CO_2 out of the atmosphere to prevent buildup and reduce its content so as to slow and eventually halt global warming and its negative impacts on society and the ecosystems that sustain them (as discussed in Sect. 4.7). This is an ambitious, yet attainable goal. However, it requires **total international cooperation** and short (a decade)/long term (four decades) investments, to a achieve a status quo condition: no additional warming, and no increase in intensity, frequency and duration of harmful weather events (discussed in Sect. 8.7).

Sanitation can be thought of having clean water to drink, toilets or engineered latrines for human excreta, and hand washing station that considered together serve to prevent disease. But diseases strike populations from other sources as cited earlier, such as biological and chemical pollution, radiation, and global warming/

climate change so that sanitation can carry extra meaning. A big question is whether nations with sanitation problems, mainly in Sub-Saharan Africa and in South Asia, will provide basic sanitation for their populations to enhance health conditions as the twenty-first century progresses [1]. As noted earlier, progress to that end is being made (see Sect. 5.2) and should continue especially as governments understand the economic, social, and political benefits good sanitation brings to nations.

Afterword

Clean water and nutritious food sustain our biological selves and proper sanitation supports public health. As healthy societies evolve they need natural resources to build shelters (homes) and infrastructure, to supply resources for manufacturing and industrial programs and to support efforts in other development sectors. Just about all countries have one or more commodities to trade whether to sustain life or spur economic development, job opportunities, and thus reduce poverty. If a natural resource or more than one is lacking and needed in one country, that country may purchase it or trade a commodity it has to offer. In this way societies establish economic, political and social relations among each other. Human resources provide physical and brain power to harvest natural resources (e.g., wood), to extract them (e.g., ores, fossil fuels), and to create (invent) uses for them as necessary for the benefit of people (e.g., renewable energy). The following chapter discusses the status of selected natural resources.

References

1. WHO and UNICEF, 2017. Progress on drinking water, sanitation and hygiene: 2017 update and SDG baselines. Online. https://www.susana.org/resources/documents/default/3-2805-7-1500888385.pdf
2. Royte, E., 2017. Nearly a billion people still defecate outdoors. Here's why. National Geographic, August.
3. UNICEF, 2018. Diarrhea caused by poor sanitation and hygiene, unsafe drinking water, remains a major cause of child malnutrition, disease and death in South Asia. Online. www.unicefrosa-progressreport.org/open-defecation.html
4. Fleming, R. and Ford, M., 2001. Human versus animals – comparison of waste properties. Online. www.ridgetownc.com/research/documents/fleming…
5. FAO-STAT, 2019. Rome. Online. www.fao.org/faostat/en
6. Enkenyan, 2019. South Korea conforms first swine fever outbreak. September 17. Online. phs.org/news/2019-09-south-korea-swine-fever…
7. Alberta, Agriculture and Forestry, 2005. Manure Composting Manual. Unpaginated. Online. www.1.agri.gov.ca/%24department/deptdocs.nsf/all/agdex8875.
8. Augustin, C. and Rahman, S., 2010. Composting animal manures. A guide the process and management of animal manure compost. North Dakota State University, 8 p. Online. www.ag.ndsu.edu/manure/documents/nm/478.pdf
9. Sanders, B., 2018. Global animal slaughter statistics and charts. Online. faunalytics.org/global-animal-…

10. Sari, O.F., Ozdemir, S., and Celebi, A., 2016. Utilization and management of poultry slaughter house wastes with new methods. Conference paper. Online. www.researchgate.net/publication/301350337
11. Malav, O.P., Birla, R., Virk, K.S., Sandhu, H.S., Mehta, N., Kumar, P. and Wagh, R.V., 2018. Safe disposal of slaughter house waste. Approaches in Poultry, Dairy & Veterinary Sciences, 2: 171-173. Online www.crimsonpublishers.com/apdv/pdf/PDV.000542.pdf
12. Bagla, P. and Kaiser, J., 1996. India's spreading health crisis draws global arsenic experts. Since, 274: 174-175
13. Dipankar, D,.Samanta, G., Mandak, B.K., Chowdhury, T.R, Chanda, C.R., Chowdhury, P.P., Basu, G.K., and Chakraborti, D., 1996. Arsenic in groundwater in six districts of West Bengal, India. Environmental Geochem. and Health, 18: 5-15.
14. Smith, A.H., Lingas, E.O. and Rahman, M., 2000. Contamination of drinking water by arsenic in Bangladesh: a public health emergency. Bulletin of the World Health Organization. Online. www.who.int/bulletin/archives/78(9)1093.pdf
15. Siegel, F.R., 2015. Cities and Mega-Cities: Problems and Solution Strategies. Springer. Briefs in Geography, 117 p.
16. Britannica, 2019. Nuclear Exclusion Zones. Online. www.britannica.com/story/nuclear-exclusion-zones
17. Digges, C., 2019. Japan may dump radioactive water from Fukushima into sea. The Maritime Executive. Online. http://www.maritime-executive.com/editorials/japan-may-dump-radio..
18. Liu, Z., Ciais, P., Deng, Z., and 42 additional authors, 2020. Near real time monitoring of global CO_2 emissions reveals the effects of the COVID-19 pandemic. Nature Communications, 11 (1): 5172-5181.

Chapter 6
Access to Natural Resources Not Water or Food 2020

6.1 Introduction

Animal and plant species survive and procreate where their ecosystems have a healthy atmosphere, clean water, nourishing food, and safe shelter. Stable ecosystems provide the capacity to support human needs and activities and will continue to do so only if there is adaptation planning to mitigate disruptive changes (e.g., by extreme weather events) that have been intensifying since the end of the twentieth century and will continue to occur as years go by in the twenty-first century. In addition to clean air, water, and food, 2020 societies rely on the Earth's ecosystems' natural resources and commodities produced from them to live a 'reasonable' quality of life. For many citizens these products have been sparingly available or financially inaccessible in the past. This has been changing slowly and steadily but socio-economic inequalities among segments of Society Earth have carried through into the twenty-first century. The hope is that these inequalities will continue to be steadily reduced for growing global populations into future decades but this is not assured, especially in Africa and Asia. Increased populations can result in a "less to go around" situation and a per capita value drop of some natural resources (e.g., water, Table 2.2). Competition for others (e.g., metals, wood) from economically advantaged nations or segments of their populations vs. the less advantaged ones will have the same result.

6.2 Human Resources

Of all the natural resources that contribute to the earth's societal carrying capacity, human resources are the most important. Clearly, education sets the foundation for success in sustainability projects. At a minimum, national aims should be to produce high school graduates, citizens with skills acquired in trade focussed schools and apprenticeship programs, and others that continue to advance their education at community colleges and/or universities. Whether from one group or another, apprenticeships or internships with mentors, whether for 1 year or more, education brings experience into what a person has to offer society. This approach is complemented by supporting programs that advance a person's field of interest and expand his/her knowledge bases into the fundamentals of other fields. This is a way to improve one's understanding of how his/her skills can best contribute to projects they are working at and maximize benefits to society. Working together, our human resources have the capability to put into play the innovativeness (creativity) that can today and in the near and far future preserve and protect other natural resources and Earth systems that allow societies to function for the good of their citizens.

To this point, I reiterate with added detail the outstanding results in Bangladesh where the government invested 25% of its development funds from 2003 to 2015 in an education program that sent 'plain-speaking' cadre to explain to people how open defecation hurt them and their families and friends health wise and economically. Cadre further demonstrated how the people could build latrines and maintain them or use other sanitation methods that would eliminate the problem. This education program was successful and brought the open defecation down from 43% of the 2003 population (~60 million people of ~139 million people) to 1% of 161 million (1.6 million) in 2015 to basically nil in the 2020 169.8 million Bangladesh population that is projected to reach 218.5 million in 2050. Sick days (and workplace transference of disease) became less, healthcare costs became less, and incomes rose.

Can similar education programs on environmental awareness (e.g., care of water, care of food, care of waste, care of ecosystems) help increase the Earth's human carrying capacity? It would seem so. Whether the increased carrying capacity will be enough to support global citizenry in 2050 is a question yet to be answered, especially for Africa, where many growing populations will likely still be suffering the negative effects of global warming/climate change (Table 6.1).

Table 6.1 World and regional populations distribution in mid-2020 and projected to mid-2035, and mid-2050, in billions [1]

	Mid-2020	Mid-2035	Mid-2050	% Change
Population world	7.78	8.94	9.88	27.0
Africa	1.338	1.897	2.560	91.3
Asia	4.626	5.112	5.331	13.2
Europe	0.747	0.744	0.729	−2.4
Latin America and Caribbean	0.651	0.724	0.759	14.2
Northern America	0.368	0.406	0.435	18.2
Oceania	0.043	0.053	0.063	46.5

6.3 Energy and Electricity

Following human natural resources in importance as a factor that can increase the Earth's carrying capacity, I place energy, with electricity being basic for serving societal needs. Table 6.2 shows the major energy sources that supported human ventures in 2017 with projections on how these are estimated to change in 2040, a generation in the future. As shown, the cleaner burning fossil fuels natural gas and biofuels show some increase and non-CO_2 emitting energy sources, especially renewable ones, show marked overall increases at the expense of declining use of coal and oil in the global effort to reduce the mass of CO_2 released to the atmosphere.

As Table 6.3 shows, hydropower contributed 2.51% to the global primary energy supply in 2017 with this contribution expected to grow to 3% by 2040. Several countries depend on hydropower as an important renewable source of electricity (Table 6.3). With global warming and consequent climate change, weather systems are in flux. Hydroelectric dams are susceptible to changes in patterns, intensity, and duration of precipitation that can cause drought and a lessening of water flow into a feeder drainage system. In countries dependent on glacial meltwater to flow to dams, the recession of glaciers over time can cause an inflow reduction and less generation of electricity. The result is that these countries have to adapt and be able

Table 6.2 The world total primary energy supplies in 2017 with projected contributions in 2040 [2]

Source	2017 % Contribution	2040	Change %
Oil	31.85	27.6	−13.3
Coal	27.12	21.5	−20.7
Natural gas	22.23	25.0	+12.5
Biofuels[a]	9.51	10.4	+9.4
Nuclear	4.92	5.5	+11.8
Hydropower	2.51	3.0	+19.5
Renewables[b]	1.83	7.0	+282.5

[a]Wood, Peat, Crop Derived
[b]Solar, Wind, Geothermal, others in development (tidal, wave, ocean thermal energy exchange)

Table 6.3 Examples of countries with a major generation of their electrical energy production in 2019 from a 'now' renewable resource: hydropower [3]

Country	Percent	Country	Percent
Norway	93.9	Canada	58.5
Sweden	44.6	Austria	66.7
New Zealand	59.4	Switzerland	51.6
Colombia	75.1	DR Congo	99.6
Venezuela	69.2	Ethiopia	93.1
Brazil	63.8	Angola	75.6

to switch to an alternate energy source for their electricity needs to be able to serve their stable or growing populations as the twenty-first century progresses. This is discussed further in the following chapter.

Energy sources vary as to where they are and how they are used depending on a national or regional access or lack of access to energy sources. For example, Saudi Arabia, a principal producer of oil, uses petroleum to serve its power needs but is nurturing solar energy production, whereas France, lacking alternate fuel sources, generates ~75% of the country's electricity using nuclear energy. Interestingly, Norway, rich in oil and natural gas but environmentally proactive, generates 96.1% of the country's electricity from hydropower. Similarly, oil-rich Venezuela has become overly dependent on hydropower that generates 73% of all its power needs. As a result, Venezuela is susceptible to droughts that reduce water flow, a condition that in recent years has caused disruptive national blackouts because of lack of planning for backup input to the country's power grid. This is ominous for the future of cities (with growing populations), and regions overly dependent on hydropower that may suffer from lack of rainfall because of global warming/climate change, a situation that has to be planned for now. This will be further discussed in Chap. 10.

In general, countries generate electricity using one or more fuels with coal, oil, natural gas, and nuclear as principal sources with hydropower and other renewables increasing in use. For heating, natural gas, oil, electricity (may be solar generated), coal and wood are used, whereas air conditioning is electricity driven. Likewise, for cooking, electricity and natural gas are the main energy providers with charcoal being used in rural regions of less developed or developing countries. The transportation sector depends on gasoline, diesel fuel, jet fuel, and electricity, whereas manufacturing/industrial operations use energy generated by, natural gas, coal, and oil.

In 2018, 20.3% of electricity generated worldwide came from renewable resources with wind and solar accounting for 9.3% and with other renewables for 11% of the total. Table 6.4 lists the countries with more than 10% of their electricity

Table 6.4 Countries that generate a significant percent of their 2019 electricity needs from wind and solar [4, adapted]

	% Electricity from wind and solar
World	8.5
Country	
Germany	28.9
Portugal	28.9
Spain	25.6
United Kingdom	23.9
New Zealand	23.5
Italy	17.3
Belgium	14.6
Chile	14.3
Romania	14.0
Turkey	13.8
Netherlands	13.3
Sweden	13.3

6.3 Energy and Electricity

Table 6.5 Comparison in percent of sources of power demand in 2018 between advanced economies and developing economies [5, adapted]

	Advanced economies	Developing economies
Industry		
Light industry	17.8	23.3
Heavy industry	14.4	26.7
Buildings		
Heating	14.4	6.7
Cooling	10.0	8.9
Appliances	27.8	17.8
Other	13.3	8.9
Transport	1.2	2.2
Agriculture	1.1	5.5

needs from renewable sources. Most of these are industrialized countries in Europe but with one each in Oceania, in the Middle East and in South America.

Energy use is distributed among light and heavy industries, transport and agriculture sectors, and residential, commercial, office and infrastructure facilities for lighting, heating, cooling, and large and small appliances. Obviously, the percent usage of these energy consumers varies with industries being the largest sector for power demand in developing economies, and buildings with the greatest power demand in developed economies (Table 6.5) [5].

The increase in energy demand for agriculture in developing economies ties in with improving their food system/food supply carrying capacities for their growing populations. Similarly, the increase in power driven industries suggests more national employment and capacity to export goods and self-produce goods to supply citizens growing needs.

In reality, there is little doubt that the global generation of power could serve a projected growing population of 9.9 billion people in 2050 or more later in the century. This is especially true for cities and mega-cities with their projected populations of more than 2/3 of the world's people by 2050. The global reserves of coal, oil, and natural gas could serve growing populations energy needs to the end of the century and into the next. However, their emissions of CO_2 when combusted only serve to fuel global warming and climate changes that disrupts today's ecosystems and citizens ways of life. This will likely continue but with diminishing effects as their use wanes. Thus, it becomes essential to increase the development and use of renewable energy sources to minimize CO_2 emission into the atmosphere. This, when coupled with CO_2 capture and use or storage projects for reducing CO_2 emissions in the atmosphere (discussed in Sects. 8.7 and 8.7.1), raises the possibility of slowing and eventually arresting global warming and ultimately bringing about a controlled reversal. Many governments honor this hope with words but not necessarily with actions because their main focus is on national/self-interest economic development above all and less so on the interests of societal/global quality of life. Since about 1988, with the formation of the IPCC and through its subsequent detailed reports, scientists from all disciplines have warned governments about the increasing threats to humanity from global warming/climate change. Certainly, the ongoing development and lowering of costs for renewable energy sources to supply electricity is encouraging and being felt in the energy market. Renewables have reduced the reliance on fossil fuels in this energy sector and is to be nurtured where environmentally appropriate (e.g., sun belts, constant wind zones). This can bring electricity to populations that lack it such as in Sub-Saharan Africa.

6.4 Metals

Metals are the raw materials of heavy industries and manufacturing plants world wide. The prime metal is iron (Fe), with many others that singly or in differing combinations are used in fabricating products. Other metals that are important to industrial/manufacturing projects include titanium (Ti), aluminum (Al), copper (Cu), lead (Pb), zinc (Zn), chromium (Cr), molybdenum (Mo), tungsten (W), manganese (Mn), lithium (Li), cobalt (Co), nickel (Ni), cadmium (Cd), mercury (Hg), silver (Ag), gold (Au), platinum (Pt), tin (Sn), arsenic (As), antimony (Sb), bismuth (Bi), uranium (U), plutonium (Pu), and the rare earths from scandium (Sc) and yttrium (Y) and from lanthanum (La) and lutetium (Lu), plus others in the periodic table of the elements. Every day we use, hear, or read about the products made using one or more metals…vehicles, airplanes, trains (including subways,), ships, farm equipment, armaments, large and small electrical appliances, tools, construction materials, communication equipment, medical diagnostic equipment and many others.

The mining industry is meeting the demands in 2020. However, it is important to know that in 2020 billions in our global populations do not have the economic power to fuel a greater demand for metal-containing products. There is little doubt that the mining sector has the capacity to increase the availability of metals to meet future needs. This depends on mining economics where a company commits to a project to explore for an ore body and evaluate one that is found. If the concentration of a commodity or more than one in an ore has a favorable existing and projected future price point, development to build an infrastructure from extraction to metal product(s), a multiyear process can be summarized as follows.

First is the exploration phase during which geologists, geochemists, and geophysicists work as teams in areas with geologic characteristics that are similar to those where mineral deposits have been found in the past. A team may study the chemistry of surface earth materials (e.g., rocks, soils, stream sediments, vegetation, water, air), plus surface measurements of the earth's magnetism and its gravity to define targets for more detailed work. This entails drilling into the earth and recovering cores of the underlying rocks to determine if the tenor (percent) of the metal commodity or more than one that may be present, is economically sufficient to justify more detailing to define the extent of a deposit in order to determine whether there is enough of the metal or metals in the subsurface to invest in a long-term

6.4 Metals

mining operation. If so, an infrastructure has to be put in place to recover rock containing the metal(s), to process it and increase the concentration of the metal(s), sending this concentrated ore to a smelter that then deliver the metal(s) to industries that prepare the metal(s) in forms the industries require for use in their products. As noted previously, one is talking years to get through the sequential stages to the end product(s).

Also, if a country that is a major source(s) of one or more metals decides for political reasons to halt or limit their exports, or do so for economic gains to drive up the price, this will stimulate countries that need the commodities to find new deposits or reopen existing deposits that were not financially feasible at a lower price. This situation existed in 2010 when China, the producer of more than 90% of the global rare earth metals, held back on the export of rare earth metals from foreign companies that used 60% of China's production of rare earth metals that were absolutely necessary for several critical products (Table 6.6). This situation stimulated mining companies to begin to increase production at other known sources or develop new ones (in California, Australia, and a few other countries), some with lower concentrations in ores that were formerly not financially feasible to go through the processes cited above the get a final product. In 2015, the World Trade Organization ruled against the Chinese action and required that they no longer limit exports. Deposits of rare earth metals continue to be developed and explored for outside of China.

Although not related to metals it is worth noting that oil producing nations kept production at levels that maintained high per barrel pricing for their benefit. The price was high and technology advanced enough to allow the extraction of oil from oil shales and go into direct competition with traditional oil extraction methods. Traditional oil producers increased production to lower the price per barrel that

Table 6.6 Examples of the products in which rare earth metals are being used [6, adapted]

Use	Rare earth elements
Wind turbines	Pr, Nd, Dy
Cordless power tools	Pr, Nd, Tb, Dy
Earphone speakers	Pr, Nd, Gd
Energy efficient light bulbs	Y, Eu
LCD and plasma screens	Y, Ce, Eu, Tb
Hybrid vehicles, magnets	Pr, Nd, Sm, Gd, Tb, Dy
Catalytic converters, cameras	La, Ce, Pr, Nd
Rechargeable batteries	La, Cd
Missile guidance, other defense	Pr, Nd, Sm, Tb, Dy
Smartphone, CD/DVD, ipod	La, Ce, Pr, Nd
Pr is praseodymium	Y is yttrium
Nd is neodymium	Eu is europium
Dy is dysprosium	Ce is cerium
Tb is terbium	Sm is samarium
Gd is gadolinium	La is lanthanum

slowed/stopped oil shale production that needed the higher price per barrel to be profitable. Initially, this worked. But the principal traditional oil producers saw their main source of government income greatly reduced with the result that they adjusted production and the price per barrel rose enough so that it allowed oil shale production to be profitable once again and restart. This see-saw game hurts both production sources. A compromise might be a price point (in 2021 = ~US$52 to US$55 per barrel) that would allow production from both.

6.5 Industrial Rocks and Minerals

Among the critical natural resources are industrial rocks and minerals. These may be defined as materials (less ores of metal, mineral fuels, or gemstones) that are useful to industries in several ways depending on their physical and chemical properties (e.g., hardness/abrasiveness, electrical conductivity, density/specific gravity, strength/compressive stress, solubility). Examples include minerals (e.g., diamond, talc), or rocks as building or decorative materials (e.g., granite, marble, limestone, sandstone), as a base for cement and lime (limestone), and as additives to concrete (sand, gravel), or their content of a non-metallic chemical element (e.g., the minerals fluorite for fluorine, apatite for phosphorus), or as a source of a metal (e.g., bauxite for aluminum, beryl as a source of beryllium). The industrial rocks and minerals number ~60 and have been well reviewed in detail for their properties, locations (geography and geology), uses, and markets [7–9].

Some industrial rocks and minerals have dual uses as a non-metallic or as a metal. For example, bauxite (a rock, not a mineral) is the principal source for the metal aluminum but also has non-metallic uses as abrasives and refractories. Also lithium from minerals such as spodumene and lepidolite is used mainly as a non-metallic in batteries, ceramic, glass, and pharmaceutical as well as a metal in cellphone and in nuclear facilities to absorb neutrons. Many components of industrial rocks and minerals may go into the manufacture of a single product such as in cellphones and smart devices (Table 6.7) [10].

6.6 Wood

Wood harvested commercially from forests is used in home building and other construction projects, in furniture fabrication, in pulp and paper and simple cardboard and corrugated boxes, in home products, and in decorative (ornamental) pieces. Unless timbering is done in a systematic and controlled way that minimizes damage to forest ecosystems, preserves areas for natural seeding, and includes reforestation as needed, the Earth's capacity to provide this commodity for growing twenty-first century populations will be limited. It may be limited as well by climate changes (rainfall and temperature) forced by global warming. Forests serve as important

6.6 Wood

Table 6.7 Example of multiple components in industrial rocks and minerals that are used in consumer products: in cellphones (and other "smart" devices) [10]

Aluminium casing: flux mineral fluorspar; refractory minerals e.g. bauxite, magnesia
Plastic back cover: filler and flame retardant minerals, e.g. talc, alumina trihydrate
Speaker: rare earth minerals
Li-ion battery minerals: lithium, graphite
Silicon chip manufacture: fused silica crucibles; silicon carbide wiresaws
Polished hi-tech screen: glass minerals e.g. alumina-silica; abrasive minerals e.g. fused alumina
Intense screen colors: rare earth minerals

sinks for CO_2 emissions and thus help damp the impact of global warming. Cutting down trees or otherwise clearing by burning swaths of forest depletes the sink capacity and hence abets the CO_2 contribution to global warming and climate change.

Timbering can be sustainable and wood available to future generations if harvesting is limited and controlled and the timbered area is replanted with saplings of the same species and time is allowed to reestablish forested areas. The ecosystems in these areas and beyond can in some cases be protected during timbering in three ways. First, by cutting vines or other entanglements connecting to other trees to prevent an adjoining tree from being torn down with a tree being felled. Second, once the trunk of a felled tree is stripped of its branches, it can be moved via a cable system or via a designated truck route for delivery to a saw mill. Dragging a tree trunk to a transport point by animals should be forbidden in order to prevent gouging and loosening soils thus making them susceptible to erosion by moving water. Lose the soil and lose the natural resources it provides. Third, clear cutting a forested site should be prohibited because it exposes a soil to erosion more so on hilly terrain. Clear cutting can also cause landslides by exposing earth material on slopes to rainwater that seeps in and destabilizes them by adding weight and decreasing friction on a slope.

Forest ecosystems are also disrupted or destroyed by government sanctioned projects to clear cut or to burning down forested areas for agriculture sector ventures (raising cattle/animal husbandry and growing money crops) as in Brazil. The burning to clear acreage for plantations to grow the raw materials for palm oil and other products has been a modus operandi in Indonesia and Malaysia. In all locations where forests are destroyed (not replanted), habitats (biomes) collapse. There is a reduction in biodiversity and the threat of extinction for some life forms as ecosystem vitality crashes. As important from a humanistic concern is the displacement of indigenous populations and small communities as has happened in areas of the Amazonian forests.

Whether production from forest raw material can increase its capacity to serve the demands of a growing number of citizens with more money to spend and desire for

Table 6.8 Change in world population living in poverty from 2002 to 2018 using 2011 PPP and $1.90/day [12]

Year	World population in billions (rounded)	Poverty population in millions	% in poverty
2018	7.6	653	8.6
2015	7.4	734	10.5
2013	7.2	804	11.2
2012	7.1	908	12.8
2011	7.0	962	13.7
2010	6.9	1,090	15.8
2008	6.8	1,228	18.0
2005	6.5	1,350	20.7
2002	6.3	1,600	25.4

wood-based products is in question. Preservation of forests and new growth of wood resources for future generations should be major goals in forestry planning as should their forests viability as CO_2 sinks and ecosystems that support human endeavors.

6.7 Population Growth, Poverty, and Consumption

With continued growth of the Earth's population by a predicted 2.3 billion more people in 2050, there will likely be a greater number of people moving out of poverty into the middle class (Table 6.8) with a corresponding greater demand for goods and service. This change will be enhanced by better educational opportunities for citizens in less developed and developing nations, more employment opportunities open to them, and better healthcare, even as their populations continue to grow. For example, in 1990 there were 505 million people in South Asia and 966 million in East Asia and the Pacific living in poverty but in 2016 the number dropped dramatically to 327 million for Asia with the result that more buying power was available to purchase goods and services thus spurring economies [11]. Basically, poverty may be defined as the level of income above which citizens can achieve an 'adequate' standard of living. Obviously, it varies with the cost of living for each country. An adequate standard of living should include, at the least, citizen access to the much cited essentials of life: safe water, nutritious food, proper housing with sanitation, and medical services. Because of hidden unemployment, countries may inadvertently set a poverty level that underestimates the real number of people living in poverty. When applying for financial assistance from international lending or granting agencies to improve a country's ability to provide the services and employment opportunities that raise people out of poverty, such an underestimation would result in lesser funding granted than if all living under the poverty level were accounted for.

As discussed earlier, natural resources may be finite in their mass and volume and limited to some populations by their geographic availability (e.g., water), whereas others (e.g., food) have the capability of increasing output to sustain growing population needs globally unless corrupted or undeliverable because of conflict or political decision. Other natural resources are renewable if managed properly (e.g., wood) but may not be able to legally increase production to meet demands of growing populations with increased incomes and desire for wood products. In this latter case, plastic substitutes will be a necessary adaptation. Some resources in the solid Earth can be considered infinite but will be available to future populations depending on their costs of extraction and production (e.g., metals, industrial minerals). Energy resources are many, varied, and readily available to world markets to serve growing populations. Some are seemingly inexhaustible but are truly finite (fossil fuels) while other are inexhaustible but subject to weather conditions that are affected by ongoing global warming/climate change (sun, wind). The tangible natural resources are complemented by intangibles that decrease stress and sustain good mental health in today's societies. These include ecosystems that provide oases for quiet time and solitude or physical activities in parks and woodlands, in the countryside, and in mountainous areas and seashores and lakes. Such ecosystems should be protected and conserved to provide the same ambiances for future more populated and densely populated societies that will surely profit mentally and physically from having them to visit, especially for urbanites.

Afterword

As the global population grows and as a demographic movement gradually builds urban populations (from 54% of world population in 2020 to 70% in 2050), there are questions about resources and public health in the decades leading to 2050. Principal among these are whether there will be water availability and food security to sustain regional populations that are projected to have high growth, especially in Africa where the population will almost double in 30 years (Table 6.1). In addition, there is the hope that existing sanitation problems in some regions will improve and thus contribute to better public health, a factor that can increase populations' contributions to their socio-economic development. There is also the question of whether governments can honestly husband natural resources in addition to water and food in order to sustain citizenry needs. Hovering over the answers to these question are the existing and projected negative effects of global warming and climate changes it causes and adaptations that can mitigate their effects. These questions will be discussed in the following four chapters.

References

1. Population Reference Bureau, 2020. World Population Data Sheet. Washington, D.C.
2. International Energy Agency, 2018. World Energy Outlook 2018. Paris, 661 p. Online. www.iea.org/reports/world-energy-outlook-2018

3. Ritchie, H. and Roser, R., 2020. Renewable Energy. Our World In Data. Online. http://ourworldindata.org-renewable-energy
4. Global Energy Statistical Yearbook 2019. Online. https://yearbook.enerdata.net/renewables-in...
5. Patel, S., 2019. 10 takeaways from the IEA's newest world energy outlook. Online. www.powermag.com/10-takeaways-from-the-iea...
6. Stratfor, 2019. The geopolitics of rare earth elements. Online.MF https://worldview.stratfor.com/article/geopolitics-rare-earth-elements
7. Kogel, J.E., Trivedi, N.C., Barker, J.M. and Krukowski, S.T., (Eds.), 2006. Industrial Minerals and Rocks: Commodities, Markets, and Uses. Soc. for Mining, Metallurgy, and Exploration, Inc., Littleton, Colorado, 1507 p. Online. www.nwbooks.com/download/industrial-rocks-minerals
8. Kuzvart, M., 2006. Industrial Minerals and Rocks in the 21st Century. In Seminarios de la Sociedad Espanola de Mineralogia: Utilizacion de Rocas y Minerales Industriales, (Garcia del Cira, M.A. y Canaveras, C., Eds.)Vol. 2, Madrid, p. 287-303. Online. www.ehu.eus/sem/seminario_pdf/SEMINARIO_SEM_2_287.pdf
9. Kuhnel, R.A., 2006. Mineral policy for industrial rocks and minerals. In Seminarios de la Sociedad Espanola de Mineralogia: Utilizacion de Rocas y Minerales Industriales, (Garcia del Cira, M.A. y Canaveras, C., Eds.)Vol. 2, Madrid, p. 15-30. Online. www.ehu.eus/sem/seminario_pdf/SEMINARIO_SEM_2_015.pdf
10. Dellgatti, C., 2018. The raw materialism iPhones and other smart devices. Cupertino Times. Online. cupertinotimes.com/rarw-materials-phones
11. Roser M. and Ortiz-Espina, E., 2017. Global Extreme Poverty. Online. ourworlddata.org/extreme-poverty
12. World Bank, 2019. Regional aggregation using 2011 PPP and $1.90/day poverty line. Online. http://research.worldbank.org/PovcalNet/index.htm?1.0

Chapter 7
Global Warming and Water 2050: More People, Yes; Less Ice, Yes; More Water, Yes; More Fresh Water, Probably; More Accessible Fresh Water?

7.1 Temperature Conditions

The warming Earth and consequent climate changes are resulting in extremes of climate and weather events that are negatively impacting human populations and other life forms. These extremes (e.g., heat waves, drought, irregular rain patterns) affected water supplies and security in 2020 and will increasingly stress supplies for growing global populations in terms of availability and access to this essential commodity for human biological needs (e.g., water for drinking, cooking, hygiene and sanitation) and for global food security. The 2019 World Water Development Report suggests that in 2050, more than 5 billion people of the projected ~9.9 billion population could suffer water stress due to increased demand, climate change, and pollution of water sources [1]. To this end, we note that the 2010–2019 decade was the hottest on record and that projections based on CO_2 and other green house gas (GHG) emissions to the atmosphere suggest that the decade 2020–2029 will be hotter yet. The CO_2 mean annual increase by decade from 1960 to 2019 presented in Table 7.1 supports the suggestion and shows that three of the highest mean annual increases by decade occurred in the last four decades. In addition, we note that 2016 and 2019 were the first and second hottest years on record (from 1880 to 2019).

The warming, forced in grand part by human activity (e.g., emission of CO_2 and other GHGs to the atmosphere), influences the hydrological cycle (Table 2.1) in many ways that affect billions of the 2020 world population in their daily needs as well as governments' socio-economic development. The availability of drinking water, water for sanitation, water for crop irrigation and animal husbandry, water for commerce and industry including transport (e.g., river barge movement of goods), water for hydroelectric dams, and reservoir dams to store water against times of drought, and water for other uses is neither available (not enough, not geographically reachable) nor economically accessible to large numbers of the 2020 human

Table 7.1 Annual mean rate of growth by decade in the atmosphere CO_2 contents in parts per million (ppm) [2, adapted]

Decade	Annual mean CO_2 growth rate (in ppm)
1960–1969	0.846
1970–1979	1.277
1980–1989	1.606
1990–1999	1.496
2000–2009	1.967
2010–2019	2.526

population. This does not portent well for ~27% more people by 2050, 30 years hence and possibly a higher number later in the century.

7.2 Fresh Water from Ice: Water Towers

Almost half of the Earth's fresh water, 3.5% of all Earth's water, is stored as ice (1.74%) in mountain glaciers, continental glaciers, and ice sheets. The mountain glaciers provide meltwater for the headwaters of major river systems and the basins they serve (e.g., the Indus, the Rhone) and are often referred to as '**water towers**' or storage locations for much of the Earth's fresh water resources (albeit as ice). The meltwater released into rivers from these "towers" sustains 300 million people living directly in the mountain areas, 1.6 billion people downstream of these areas, and ecosystems nourished by the rivers. The 1.9 billion people served by melting glaciers in 2020 represent ~25% of the world population. Melting glaciers also provide water for land-based animal and plant life in ecosystems worldwide and water that discharges into estuaries and oceans carrying nutrients that contribute to their estuarine and marine food webs and food fish they harbor.

There are four principal global water towers: the Himalayas, the Alps, the Andes, and the Rockies. Regionally, within these areas and others at high altitudes or high latitudes worldwide, there are thousands of smaller water towers. A recent assessment of the vulnerability of water towers in terms of their role of supplying water to downstream societies and ecosystems dependent on their flow was published after a study of 78 water towers and the basins they service [3].

Meltwater that flows from mountain glaciers has not yet hit peak water discharge in some regions (e.g., in North America and Asia) whereas in others (e.g., the Alps and Andes) meltwater discharge has peaked, is in decline, and feeds less water to streams, rivers and glacial lakes for human use, to ecosystems, and to recharge unconfined aquifers [4].

Water tower vulnerability has been assessed in terms of populations' water stress, governance (management/adaptation), hydro-political tension (serving more than one province/country), contemporary and future socio-economic changes (**more people, uneven economic status**), and future climate changes (**continued global warming**). Clearly, these changes are most important in a population water vulnerability assessment. The water towers report focused on the number of people affected by each in four important regions of the continents affected (Table 7.2).

7.2 Fresh Water from Ice: Water Towers

Table 7.2 Examples of people dependent on meltwater from mountain glaciers that feed streams, rivers and lakes. For the Alps and Andes, the flow of meltwaters is on the decline (past peak discharge). Glacier meltwater discharge from the Himalayas and Rockies has not peaked and is forecast to continue as usual or increase until mid-century [3, adapted]

Past "Peak Discharge"
Europe: **Rhone**—serves 10 million people in France, Italy and Switzerland
Rhine—serves 41 million people in France, Germany, Austria, Italy, Switzerland, Netherlands, Liechtenstein
South America: **South Chile, Pacific Coast**—serves 7 million people in Chile and Argentina
Negro—serves 1 million people in Chile and Argentina
Not Yet At "Peak Discharge" (projected for 2050)
Asia: **Indus**—serves 235 million people in India, China, Pakistan, Afghanistan, Nepal, and Kashmir
Ganges-Brahmaputra—serves 487 million people in India, China, Bhutan, Myanmar, Bangladesh, Nepal
North America: **Columbia and NW United States**—serves 9 million people in US (Washington, Idaho, Montana, California, Oregon, Wyoming) and Canada (Alberta, British Columbia)
Fraser—serves 3 million people in United States (Washington) and Canada (British Columbia and Alberta)

It should be noted that 722 million people, or ~38% of the 1.9 billion people dependent on availability of and access to glacial meltwater, live in the Asian Indus and Brahmaputra-Ganges basins listed in Table 7.2 [5].

Lastly, the cited report counsels the obvious. Governments should talk now to format climate change adaptation policies that safeguard populations from depleting **seasonal** water supplies in the mountain regions and also maintain a satisfactory water flow for use and sustenance of societies farther downstream [3]. Talks should be followed with urgency for adaptation planning and actions.

7.2.1 Some Consequences of Declining Meltwater Flow from the Alps

Many mountain glaciers have not peaked yet in the volume of melt water they discharge so that people in their drainage basins go on with life as usual. However, their futures are at risk as illustrated by some basins where ice has thinned sufficiently with melting and reduced river flow has been measured. For example, in Switzerland, a country that derives 83% its fresh water supplies from aquifers, river flows have been reduced by 10% since 2014 resulting in less recharge of the aquifers. This also presents a long-term risk for hydroelectric power. In Switzerland, the 60% energy needs satisfied by hydropower in 2020 are forecast to fall to 46% by 2035 as a result of less precipitation in the mountain glaciers regions and an increase in national energy use. Overall, in Europe 20% of electricity comes from hydropower but this percentage is expected to decrease in coming decades as the result of a flow reduction in the headwaters of great rivers (e.g., the Rhone, the Rhine, others) fed by glaciers in Switzerland (and in other countries). Most of the Rhine glacier field may be gone by 2100 if global warming continues at its present rate or sooner if the rate of warming accelerates.

A less emphasized threat to a decline in flow (and warming) of European rivers has to be the use of river waters and canal waters drawn from them as coolants for the continent's nuclear power facilities reactors that exist or that are to be decommissioned in the future (Fig. 7.1) [6].

In the past, nuclear reactors have been temporarily shut down during heat waves when sources of coolant waters were warmed too much for them to be drawn into reactor systems. What adaptation(s) do the European Union members and other countries have in place for potential problems of declining water flow and/or water that is too warm seasonally for use as reactor coolants?

During the European summer of 2019, especially in July, a combination of an extreme heat wave and severe drought caused France to shut down two nuclear reactors in SW France and curtailed the output of six other reactors in order not to break the environmental limits of the temperatures of rivers at low flow that is used for cooling water. This reduced electricity output by 5.2 GW (an 8% reduction) at the time electricity demand for air conditioning systems increased. Germany shut down one reactor for the same reason [7]. Operators of European nuclear plants at risk of warm water or low water flow to reactors that can cause reduced output or shut down have to evaluate now whether their backup cooling water sources that can be piped to reactors in order to sustain electricity output at normal or reduced levels

Fig. 7.1 Position of nuclear power plants in Europe (2011). With 85 nuclear power plants and 159 reactors in operation, these represent one third of the global fleet [8]

will be a viable adaptation. Planned nuclear power plants in Europe and worldwide have to consider if their sites are vulnerable to climate events that could affect the required water supply during the expected life of a facility, and as needed, alter plans accordingly…new site, or available of secondary cooling water or low water replacement sources.

7.2.2 Some Consequences of Declining Meltwater Flow from the Andes

A 2013 report from the European Geosciences Union pointed out that glaciers in the tropical Andes have shrunk by an average of 30–50% since the 1970s as regional temperatures increased by an average of 0.15 °C per decade. The melting has been greater since the 1970s for small glaciers at low altitudes (<5400 m) or at twice the rate of high altitude glaciers. This is with little change in rainfall so as not to account for the glacial shrinkage. Inhabitants of Peru, Ecuador, Bolivia, and other tropical countries plus populations in southern Chile and southern Argentina rely on **seasonal glacial meltwater for continued river flow** for drinking, and/or agriculture (for agrobiodiversity and food security), and/or hydropower that is critical now in the countries cited above plus Colombia. Governments have to institute projects now that ease or eliminate any water shortages that will surely affect their post-2020 populations way of life and that is likely to affect them more so approaching mid-century and later in the century [9]. The European planning to make up for water lost via declining mountain glacier melt is less of a problem compared to South America because of Europe's contracting populations that are projected to continue into the future for many nations. However, planning to deal with the loss of mountain glacier meltwater discharge into streams, rivers and glacial lakes will be more complex in South America. Here, national populations are projected to continue to increase from 430 million people in 2020 to 491 million people in 2050 so that the ~19 million inhabitants impacted in 2020 by declining glacial meltwater flow can be expected to increase in numbers by mid-century [10]. Except for landlocked Bolivia and Paraguay, desalination will undoubtedly be an adaptation used to make up for some of the water loss by the continent's coastal countries. Interestingly, the loss of glacial meltwater may ease somewhat because of a South American population that is projected to contract during the latter half of the century to 429 million people.

7.2.3 Glacial Ice Thinning

The fact is that the melting-ice thinning process will affect more and more of the world's low- and mid-altitude mountain glaciers as global warming continues to increase beyond the 2020 1^+ °C mean rise above the pre-industrial temperature. To the present, the meltwater volume from high-altitude glaciers has not been affected so as to reduce their water flow, but as global warming continues there will be an increase in the rate of loss in ice thickness and areal extent of these glaciers as they recede. As Himalayan region glaciers rate of meltwater discharge declines in the future (2050?), water and power needs for developing economies such as those in India (e.g., Ganges-Brahmaputra region), China (Yangtze region), and the Mekong regions will be increasingly at risk. The gradual loss of meltwater seasonal discharge that regions depend on for continued river flow that sustains population needs during uneven spring and summer rainfall will impact the quality of life for 100s of millions in less developed, developing and developed nations. The impacts may be eased if alternate water source plans are proactively put into place and tested before a need intensifies rather than try to "play catch up" later. Economic efforts to resolve future water problems during the 2020 decade will cost far less than striving to resolve them future decades (see Sect. 11.1, second and third paragraphs).

7.3 Changes in the Water Supply/Security Exclusive of Mountain Glaciers

As discussed in previous paragraphs, the volume of fresh water released from mountain glaciers to river basins is on the decline in much of the Alps and Andes and is projected to pass peak discharge in much of the Himalayas and Rockies during mid-twenty-first century. Glacial meltwater discharges into drainage basins and presently makes up for the **lesser contribution of rainwater** to streams, rivers, and lakes **during the late spring and summer**. The meltwater thus maintains a 'normal' water flow in a basin. This meltwater input has been lost in some regions and will be lost in many others as global warming continues throughout the century. Planning must consider how to make up for the lost spring/summer glacial meltwater. One answer is to store sufficient rainwater in reservoirs (dams) for release as needed during spring and summer shortfalls of precipitation in a basin. Proposed locations for the siting of such purposed reservoirs is critical from physical (geological conditions), biological (disruption of upstream and downstream ecosystems/life form activities), and social aspects (displacement of established populations) [11].

The transference of fresh water from mountain glaciers that are receding or will recede in the foreseeable future (2050?) to the oceans through drainage basins and by calving ice from continental glaciers and broken off sections of ice sheets that are undermined in the warming seas affects the global hydrological cycle. Will the amount of water for human use be less or more or remain the same is a question yet to be answered. What we do know is that as a result of global warming, melting ice masses and thermal expansion of seawater causes sea level rise. Rising sea level increases oceans' surface areas as seawater encroaches onto coastal zones. As a result of warmer seawater and an increasing ocean surface area, more moisture evaporates into the atmosphere. Researchers determined that water vapor increases by 7% for each 1 °C of warming. This suggests that the total worldwide volume of precipitation will likely increase by 1–2% per degree centigrade rise with precipitation increase in some regions but with decrease in others [12].

With an expected increase in the Earth's yearly precipitation through the twenty-first century and projected regional differences in seasonal precipitation volumes and intensities, there are added risks to water security of vulnerable populations/locations. The degree of risk depends on rate and magnitude of warming, geographic location (e.g., topography [mountain ranges, proximity to sea coasts]), level of development (social and economic), and feasible options to adapt and mitigate threats (political decisions). Such factors affect runoff into waterways, aquifer recharge rates, and hence a population's ability to capture and access additional water resources. The populations most at risk of a lessening of water security with passing decades are in Asia and Africa and to a lesser degree in South America. **Additional terrestrial precipitation and fresh water capture from the aforementioned increase in evaporation from expanded ocean surfaces will likely add to the recharge of unconfined aquifers and add to a planned 'capture and storage' of safe water.** Whether this increase in safe water supply can meet the earth's human carrying capacity needs for another ~2.3 billion more people on earth in 2050 or a possible 1.1 billion additional people by 2100 is questionable. This is in question because regional deficits in water supplies in 2020 fails 1 billion people with no ready access to safe drinking water and 2 billion without proper sanitation.

7.4 Future Regional Water Availability (Increase, Stable, Decrease) with Progressive Global Warming Through the Twenty-First Century

Reports from the IPCC and other organizations and researchers 'computer calculate' that regions that are wet are likely to get wetter from increased precipitation and greater precipitation intensity from high energy tropical storms especially in high latitude regions. How this affects specific locations within the regions is uncertain. Conversely, dry regions of the sub-tropics, the so-called 'horse latitudes' between about 30° to 35° N and S (in mid-latitudes), areas of calm winds and with high pressure that suppresses precipitation and cloud formation, will probably get drier and expand towards the poles. Latitudinal divisions are given in Table 7.3. In a region with a more restricted latitudinal scale such as Europe, computer generated models suggest that much of the continent will likely have wetter winters but that central and southern regions will to have drier summers [12].

Each country within these latitudinal regions can be divided into 5 general climatic regions and 30 specific Köppen climate zones. The climatic regions are

Table 7.3 Traditional latitudinal divisions for high, middle, and low latitudes and regions that fall in them

High Latitudes Regions: From 66° 33′ 39″ to 90° N and 66° 33′ 39″ to 90° S
Includes Siberia (Russia), Northwest Territories (Canada), Greenland, Scandinavia, Alaska (United States)
Mid-Latitudes: Regions Between 23° 26′ 22″ and 66° 33′ 39″N and 23° 26′ 22″ and 66° 33′ 39″S
Includes parts of North America, South America, Africa, Asia and Australia
Low Latitudes: Regions From Equator 0 to 23° 26′ 22″N and 0 to 23° 26′ 22″S
Includes Northern South America, Central America, central and Saharan Africa, South and Southeast Asia

tropical (with 4 climate zones), dry (arid, with 4), temperate (with 9), continental (cold, with 11), and polar (with 2). The five general climate regions and each one's climate zones have been classified on the basis of precipitation type and level of warming [13, 14].

Of 201 countries listed according to the latitude of their capitals, 97 are in the mid-latitudes, 104 in the low latitudes, and none in the high latitudes. However, away from the capitals, nations may contain multiple Köppen climate zones latitudinally north to south and longitudinally east to west depending on a nation's size and attitude, influence of topography by elevations above mean sea level, orientation with respect to the sun, and proximity to water bodies. The United States and China have the highest number of zones at 18, followed by India with 16, Argentina, Mexico and Peru with 13, Bolivia with 12, South Africa with 11, and Australia, Colombia, Ecuador, Nepal, Russia and Turkey with 10. Fewer Köppen climate zones are found, for example, in countries in the (arid) Middle East (e.g., Israel and Iraq with 4 and Syria and Iran with 5) and in (cold) Scandinavia (e.g., Sweden and Finland with 2 and Norway with 3), and in (humid) SE Asia and Africa (Thailand, Vietnam, and Indonesia with 3) and Nigeria with 4 and DR Congo with 5).

Figure 7.2 presents a map on which "fixed" latitudes establish the referenced low, middle, and high latitude divisions. However, on the basis of satellite analyses since 1979, researchers have shown that the edge of the tropics is expanding about 0.2° to 0.3° latitude per decade. After ~4 decades (1979-to-2018) the shift approximates 1 °C latitude north and 1 °C latitude south or ~69 miles (~110 km) north and south. As the sub-tropics zone expands poleward so do extensive dry areas in the tropics. Thus, for example, the boundary between where it is getting wetter and where it is getting drier is pushing farther north making Germany and Great Britain less wet and the already dry Mediterranean region drier yet. As this process evolves, the smaller equatorial areas with heavy precipitation are predicted to contract. The gradual shift of the sub-tropics and tropics due to global warming-forced climate changes is expected to exacerbate water availability and limit the human carrying capacity for growing populations in existing water poor regions of Africa, South America, and the Oceania and equatorial Asian regions [15].

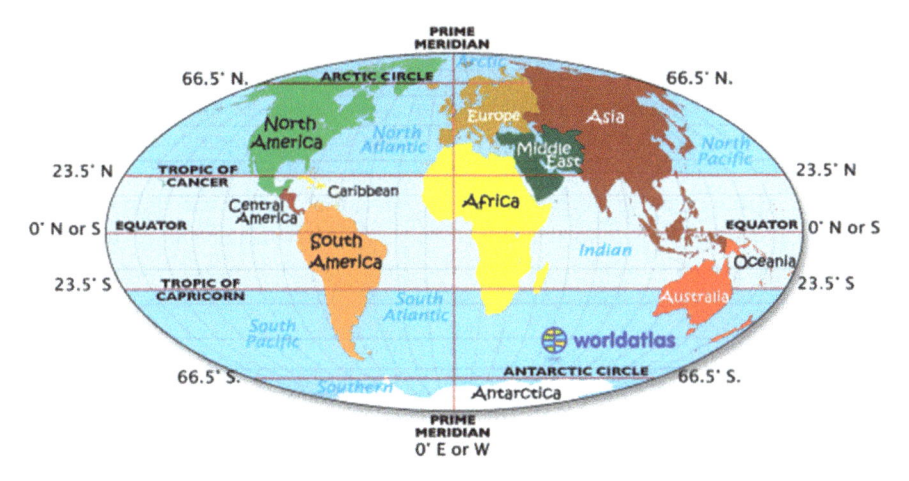

Fig. 7.2 Latitude zone map. Note: the subtropics division, 23° to 35° N and S of the equator are not shown [16]

7.5 Water at Risk/Water Availability/Water Food/Water Hazards as Twenty-First Century Progresses

The availability of water to serve society as the twenty-first century progresses is dependent on multiple previously cited factors. A principal factor is growing global populations (especially in Africa and Asia) and demographic shifts by rural citizens to growing urban centers where water availability to meet societal needs for some is already at close to its limit or scarce. The World Bank calculated the impact of water deficiency on the 2016 7.5 billion global population as 2.2 billion people without safely managed drinking water services, 4.2 billion people without safely managed sanitation services, and 3 billion without basic hand washing facilities that if corrected can prevent human to human transfer of bacterial or viral diseases [17]. The hand washing problem has improved markedly during the past few years but still presents a public health problem. If we consider these numbers as likely representative of the conditions for the 2010–2019 populations, then **what can we expect** for the 2050 global population? One expectation is the same relative numbers lacking sufficient water for ~27% more people on Earth **if little or nothing is done to improve water availability** (2.8, 5.3, and 3.9 billion without safe drinking water, without better sanitation, and without more hand washing stations, respectively). Another is **an improvement in water management** that favors reduction in water stress. A third is **increased water stress** to the point that there are major migrations driven by a society's sense of survival to where water is available. In all scenarios, water stress has more of an impact on the vulnerable poor sectors of populations globally, mainly outside of Europe, North America, and parts of Asia.

As already noted, the World Bank reported that water is essential to food security with 70% of available water used in agriculture (especially for irrigation) in 2020

[17]. The FAO estimates that food production will have to grow 69% to feed an increased 2050 population but without an added draw on an existing water supply that also serves domestic and industrial sectors. This can be accomplished by innovative water management following agro-technical advances (hybridization, genetic modification, adaptations as drip irrigation) that can reduce water needs/waste without affecting crop yield and quality [18]. The water thus 'saved' can support conditions for growing populations that in part help fulfill UN sustainable development goals: improve public health and also allow continued economic development that promotes gainful employment and thus reduce poverty.

7.5.1 Threats to Coastal Cities and Aquifers

In several urban centers with growing populations water is being pumped from aquifers at a greater rate than it is being recharged so as to ultimately deplete this life sustaining resource (e.g., in South and Southeast Asia). With this forewarning, nations must proactively develop the infrastructure to use surface waters where feasible. In addition, the aquifer discharge is reducing the buoyancy pressure in an aquifer thus resulting in **subsidence** of the land surface that makes coastal urban center populations more vulnerable to **tidal flooding** and **hurricane driven storm surges** abetted **by** the global warming forced rise in sea level. Computer models show that by 2100 at an Earth temperature of 1.5 °C over a pre-industrial temperature, sea level would rise 0.26–0.77 m (~10–30 in.). For a temperature of 2 °C, the sea level rise could be 0.36–0.87 m (~14–34 in.) [19]. In either scenario, there would be a **higher pressure of a sea water wedge pushing landward to invade a coastal fresh water aquifer** where the aquifer rock formation intersects the continental slope. This would diminish the coastal aquifer freshwater supply unless at risk cities continue to adapt to the situation by management efforts that stop a seawater invasion and protect the freshwater supply [20].

A principal factor that affects the availability of water is the uneven distribution of water resources geographically. There are locations of "plenty" and locations of "deficiency" that are not in reasonable proximity to allow ready transference of water from one to another. **The 'reasonable proximity' distance depends on whether global investments are made to build pipelines to move water from where it is plentiful to where it can relieve stress at water deficit locations.** As noted in Chap. 2, Saudi Arabia has built 4000 km of pipeline to carry potable water from its many desalination plants on the Red Sea to the country's population centers. An "elephant in the room" is the increasing CO_2 content in the atmosphere that is driving global warming in 2000 from 1+ °C above the preindustrial temperature to 1.5 °C above it. This could likely be by 2030–2052, at the current rate of increase 0.15 °C decade (0.1–0.3 °C range) by which time planned adaptations that mitigate impacts on water poor societies (e.g., availability and food security) should be in place. As noted earlier in this chapter, for a global warming mean temperature increase that reaches 1.5 °C above a pre-industrial level, computer models forecast an increase in extreme temperatures in many regions, an increase in intensity of and amount of heavy precipitation in others, and an increase in intensity and duration of drought in still others [21].

The IPCC, the World Resources Institute (WRI), the UN, the World Bank and other reporting groups identify regions where climate change conditions today and those projected for the future (e.g., to 2050) would affect water security as demand increases from population, agricultural, and industrial pressures [22, 23]. Any significant reduction in precipitation in these regions will affect the water supply that supports human domestic needs as well as the volume of water needed by the 16% of food crops fed by irrigation or for irrigation of rain fed crops when there is an extended drought. In essence, the greatest climate driven deficits in water supply/security for food production are projected to be in and around the mid-latitude regions (i.e., between ~30° to 35° N and S of the equator in the 'horse latitudes'). Several regions are projected to become drier and most vulnerable to water stress as the earth continues warming and climates change. These include the Middle East (including northern Africa), Sub-Saharan Africa, the Mediterranean region, western North America (USA [southern Texas], Mexico), western Asia, and northern China in the northern hemisphere, Chile, Argentina, and South Africa, in the southern hemisphere, and eastern and southeastern Australia. The degree of "drier" for 'local' areas within a region is difficult to predict with existing climate models. In many of these regions, countries are in development modes, populations are expanding, and there is an increasing demand for access to the "single water pie" with a fixed amount of water available for food production, to service domestic needs, and for energy and other sectors.

7.5.2 Rainfall Deficit Impact on Hydropower

As noted earlier (Sect. 6.3), deficits of precipitation can affect a population's electricity grid as hydropower has to be cut back because a reservoir water level becomes too low so that electricity may be limited to specific hours during the day. In 2014–2015, a severe drought in Brazil reduced the hydropower source at the Itaipu Dam (latitude 25°32′52″S) that decreased the electricity available to citizens of the mega-city Sao Paulo. In 2010, 2016, and 2019, Venezuela rationed electricity because lack of rainfall affected power generation at the Guri Dam (location 7°45′N). In southern Africa in 2019, Zimbabwe and Zambia have had to cut back on electricity use (Zambia) or ration electricity (Zimbabwe) because of low water in the reservoir of the Kariba Dam (latitude 16°31′S) as a result of drought that reduced the Zambezi River to 25% of its normal flow. The latter two locations are in low latitude regions and Itaipu is basically in an edge zone where a low and mid-latitude zone blend together.

7.5.3 The Ethiopian Dam Threat to Egypt's Water Lifeline: 2021 Conflict Resolution Necessary

The Grand Renaissance Ethiopian Dam is being built on the Blue Nile to supply all electrical energy needs for urban and rural Ethiopia. It will be the largest hydropower facility in Africa and seventh largest globally. Any excess power generated could be exported to Sudan, Egypt, or Djibouti. The dam poses some problems for the Sudanese water supply but is a threat to population survival for Egypt. The Nile provides 95% of Egypt's fresh water supply for 95 million Egyptians that live in the Nile Valley and for irrigation of farmland that produces 50% of the food supply for all of the country's 2020 100 million population. Eighty-five percent of the Nile flow in Egypt originates in Ethiopia (the Blue Nile) and 15% in Sudan (the White Nile). The rate of filling of the dam reservoir proposed by Ethiopia threatens Egypt's water lifeline that serves the domestic and agricultural needs for it's 100 million (and growing) population. Ethiopia wishes to fill the dam in 3 years while Egypt wants it to be filled in 15 years to allow time to prepare for the future (e.g., construction of desalination plants). Any significant permanent reduction in the Nile river flow into Egypt water lifeline as the dam is begins to fill and thereafter is a threat to the water and food security of the Egyptian population. It would reduce recharge of aquifers and also diminish electricity production from the turbines of the Aswan Dam. The two countries and Sudan are in discussion to find a solution amenable to all parties. A major problem to be resolved is the volume of water in the reservoir allotted to Egypt and Sudan during times of drought that are expected to increase with the rising global temperature. During July, 2020, the reservoir began to fill. If the Nile flow into Egypt drops to less than necessary fill the water needs of its population, conflict could result. The UN, the African Union, and other parties and their technical advisors are working to solve the problem and avoid a destructive confrontation in which all parties would be losers.

Thousands of new dams are in a planning stage or in construction worldwide, some for flood control, some as reservoirs, and others for renewable "green" hydropower [24]. The effects of a planned dam on water flow to abutting regions or downstream countries and water security in the twenty-first century future has to be planned for along parallel tracks: (1) with respect to the effects of climate change and estimated river flow, and (2) to reduce conflicts of interest with respect to water demands of growing populations and their basic needs for safe water, water for food production, and water for manufacturing and industry.

7.6 A Solution to Bring Water to Water-Starved Populations

Globally, in 2014 the world had 3,500,000 km (2,175,000 mi) of pipeline to transport natural gas, petroleum, and petroleum products from points of production to points of processing, and on to points of sale or use, ofttimes over thousands of kilometer/miles (see Sects. 2.5.1 and 2.5.2). The United States has the most extensive pipeline system with 548,665 km (340,785 mi) to move natural gas and 244,260 km (151,714 mi) to carry petroleum and petroleum products [25]. Nationally, some countries have made and are making investments in pipelines and river diversions to transport water from distant or relatively nearby sources to cities in arid regions and to arid regions that a country wishes to develop. For example, Libya (pre-conflict, oil rich) has its Grand Manmade River that supplies water to Tripoli, Bengazi, Sirte, and several other cities through 2800 km (1700 mi) of pipeline from 1300 wells that pump water from a 500 m (1600 ft) deep aquifer. China is diverting water thousands of kilometers from the Yangtze and other rivers in the south to water needy cities and agricultural/industrial projects in the north with reaches into China's northwest Tarim Basin for planned development.

There is no question in my mind that in an attempt to resolve the planet's problem of uneven geographic distribution of water and the growing global population in many water poor areas of Africa, Asia, and South America, there is a lone solution, an international one. This is the construction of pipelines that can carry water from areas with surpluses of clean water to those areas in desperate need of safe water. With data on existing sources of water and societal needs, plus projections of how these are likely to change in the future because of global warming, climate changes, and population growth and migration to cities and mega-cities, planning is essential now. Implementation of the plans for construction of pipeline infrastructures sooner rather than later can mitigate or prevent water driven humanitarian catastrophes in the future.

As of 2014, 118,623 miles (189,797 km) of pipeline for hydrocarbon industries are either in the planning and design stage (75% or 88,697 mi [141,915 km]) or are in the construction stage (25% or 29,656 mi [47,450 km]) [26]. I ask myself how many people in the world population today or in the future could have clean water available and accessible if there was funding for only the in construction stage 29,656 mi (47,450 km) of pipeline planned to transport oil to instead bring water to people in critical water deficient locations from where it is now plentiful and projected to continue be plentiful during the century. My estimate is 1/4 of the Earth's population.

The cost to construct pipelines varies depending on several factors but runs into million of dollars or euros per linear mile. In general terms, the cost may be broken down into three or four major expenditures. These are **materials** (for pipes [varying widths]) and **labor** (for excavation, inspection, setting pipe in place, removal of excavated earth materials, environmental assessment), **professional services (for surveyors and engineers)**, nature's challenges (dealing with topography), and where necessary **right of way costs**. Like any major project there will be accountants and lawyers to monitor a working budget. Where might the funding come from is the challenge that may be the major problem. A starting point is with the countries themselves that need the water to step forward initially with a tranche of self funding and then present proposals to financial institutions for loans or grants. Sources for such infrastructure funding are the World Bank, the European Bank for Reconstruction and Development and regional banks (the African Development

Bank, the Asian Bank, the Asian Infrastructure Investment Bank, and the Inter-American Development Bank). In addition, funding is possible as grants from agencies in developed countries with expected post-COVID-19 pandemic strong economies and monetary reserves (e.g., the United States, the European Union, the United Kingdom, France, Germany, the Netherlands, Norway, Japan, China, and others). The United Nations Development Program can provide technical expertise to complement national expertise and that from well-reputed global consultants. As with all projects designed to adapt to global population growth and predicted impacts of global warming and changes in climate, it would be more beneficial to invest in working to solve water problems sooner rather than later. Adaptation programs take time to become operational and as time passes costs will increase greatly even with adjustments for inflation and there will be increased losses/damage incurred to people, property, and ecosystems in the interim years between non-action and action.

Afterword

Ultimately, food production to serve growing global populations is dependent on water and fertile (nutrient rich) soils and sun. As such it is hostage to climate changes such as increasing temperatures and water availability, whether farmland is rain fed (80%) or irrigated (20%), and food animals being watered and fed. The following chapter evaluates the future regional/global food supplies under an increasing global temperature, changes in precipitation patterns, and of course, growing global populations. It further reviews how dietary changes, if implemented, could feed a projected 2050 population of up to ten billion people. The chapter revisits the driving force of global warming (emissions of CO_2 and other GHGs) and offers commentary on what is being done in 2020 and what technologies are being researched in pilot programs on additional ways to slow and ideally stop global warming during the next three to four decades.

References

1. UNESCO, 2019. The United Nations World Water Development Report 2019: no one left behind. 11 Chapters, 186 p., Paris. Online. https://en.unesco.org/publications/world-water
2. NOAA, Earth System Research Laboratory, Global Monitoring Division, 2020. Online. www.esrl.noaa.gov/gmd/ccgg/trends C. D. Keeling, Scripps Institute of Oceanography
3. Immerzeel, W. W., and 31 co-authors, 2019. Importance and vulnerability of the world's water towers. Nature, 9 Dec. Online. https://doi.org/10.1038/c41586-019-1822-y
4. Huss, M. and Hock, R., 2018. Global-scale hydrological response to future glacier mass loss. Nature Climate Change, 8: 135-140.
5. Borunda, A., 2019. The world's supply offers water in trouble as mountain ice vanishes. National Geographic, December. Online. www.nationalgeographic.com/science/2019/12/water…

References

6. Thomasson, E., Hydropower industry braces for glacier-free future. Online. www.reuters.com/article/us-climate-glaciers/…
7. Reuters Staff, 2019 (July 25). Hot weather cuts French, German nuclear power output. Online. www.reuters.com/article/us-french-electricity…
8. Nuclear Transparency Watch, 2014. Position of the nuclear power plants in Europe. Online. www.nuclear-transparency-watch.eu/documentation/position…
9. Rabatel, A. and 27 co-authors, 2013. Current state of glaciers in the tropical Andes: a multi-country perspective on glacier evolution and climate change, Cryosphere 7: 81-102. Online. www.the-cryosphere.net/recent_papers.html
10. Westbrook, J., 2017. The effect of melting glaciers on tropical communities. From statement by Professor K. S. Zimmerer re Peru, Bolivia, Colombia and other western and northern South American communities. Online. www.sustainability.psu.edu/spotlight/effect…
11. Siegel, F. R., 2016. Mitigation of Dangers from Natural and Anthropogenic Hazards: Prediction, Prevention, Preparedness. Springer Briefs In Environmental Science, 127 p.
12. IPCC, 2014, AR5 Climate Change 2014; Impacts, Adaptation and Vulnerability. Chapter 23. Europe, pp. 1269-1325. Online. www.ipcc.ch/report/ar5/wg2
13. Köppen, W., 1936. Das Geographische System der Klimate. In Köppen, W. and Geiger, R. (Eds.), Handbuch der Klimatologie, Vol. I, Part C, Gebrüder Borntraeger, Berlin.
14. The Köppen Climate Classification. Online. https://zh.mindat.org/climate.php
15. Staten, P.W., Lu, J., Grise, K.M., Davis, S.M. and Birner, T., 2018. Re-examining tropical expansion. Nature Climate Change, 8: 768-775.
16. www.worldatlas.com/atlas/images.htm
17. World Bank, 2019. Water. Online. www.worldbank.org/topic/water/overview
18. UNFAO, 2009. 2050: A Third More Mouths To Feed. Online. www.fao.org/news/story/en/item/35571
19. IPCC, 2019. Special Report on the Ocean and Cryosphere in a Changing Climate (SROCC). Unpaginated, 6 Chapters. Online. www.ipcc.ch/srocc
20. Siegel, F.R., 2019. Adaptations of Coastal Cities to Global Warming, Sea Level Rise, Climate Change and Endemic Hazards. Springer Briefs in Environmental Science, 86 p.
21. IPCC, 2018. Global Warming of 1.5°C (SR1.5). 5 Chapters. 538 pp. Online. www.ipcc.ch/sr15/download
22. IPCC, 2019. Special Report on Climate Change and Land. Summary for Policy Makers. 7 Chapters. Online. www.ipcc.ch/reports
23. Maddocks, A., Otto, B. and Luo, T., 2016. The Future of Fresh Water. World Resources Institute, Washington, DC. Online. www.wri.org/blog/2016/06/future-fresh-water
24. Zarfl C., Lumsdon, A.E., Berkekamp, J., Tydecks, I. and Tochner, K., 2015. A global boom in hydropower dam construction. Aquatic Sciences, 77: 458-462.
25. CIA World Factbook 2016. Online. www.cia.gov
26. Wikipedia. Pipeline transport. Extracted from Pipeline and Gas Journal, 2014. Online https://en.wikipedia.org/wiki/Pipeline_transport

Chapter 8
Food 2050: More Mouths to Feed—Food Availability and Access

8.1 Introduction

Food production and food security depend primarily on cropland under cultivation in nutrient/micro-nutrient sufficient soil, water to support agriculture (food crops, livestock), and daily hours of sunlight exposure (warmth and photosynthesis). Second, food production and food security are enhanced by crops that are or can be made pest resistant, disease resistant, and less affected by the use of herbicides for weed control. Third, biodiversity of agricultural land supports food productive soils and should be protected and preserved. Fourth, cropland should not be increased by encroaching into forest terrain or other viable ecosystems housing natural resources that support societal needs other than food. The aim should be to increase yield and quality of food products to the maximum on existing farmland and arable land approved for cropping. Improved management can achieve that end. This includes monitoring critical factors and advising how achievable improvements for a given environment can increase crop yield and quality for production and delivery to consumers. These factors include soil fertility, nutrient replenishment, efficient water use, planting of crops that best serve human food needs and that are best adapted to environmental conditions. Complementing the factors are crop protection as cited above, silo storage and transport container adaptations (e.g., hermetic sealing, refrigeration) that protect products from loss to contamination and animals, and minimize food waste (in homes, markets, and restaurants). Food security of vulnerable populations is being enhanced in some countries/municipalities by the daily collection and delivery of unsaleable but safe and nutritious foods from markets and unused comestibles from restaurants to food banks to feed needy citizens and their families (see Sect. 3.5.6, fourth paragraph). During 2020 and into 2021, a surge in citizen food need is the result of a nation's economic downturn and people's loss of employment and income because of the 2020/2021 COVID-19 pandemic.

8.2 The Challenges

Whether changes to increase food production and improve food distribution as cited above can nourish an expanding global population that is expected to grow from 7.8 billion people in 2020 to ~9.9 billion in 2050 is in question if we assess 2020 realities, and less probable for populations later in the century. The realities include the already cited 821 million people with malnutrition and 2 billion micro-nutrient deficient persons in 2020. Of great concern in these case are 151 million children globally stunted in growth and possibly cognitively impaired, and 51 million children wasted (low weight for height), all signs of malnutrition. At present, enough food is produced worldwide to feed the global population but lack of storage against loss, insufficient and inefficient transportation infrastructure and distribution networks, bad government decisions, and dangers in war/conflict zones prevented delivery or distribution of life sustaining foodstuffs for food starved populations. Even if these problems were eliminated, it would still be necessary for the previously quoted FAO and World Bank estimated increase of ~70% in food supplies to feed an additional ~2.3 billion more people in 2050 than in 2020. Given water supply limitations and acreage devoted to animal feed and biofuels, changes would have to be made in food systems from production to consumption in coming decades to meaningfully reduce global undernutrition, malnutrition and famine conditions. Positive changes in a population's nutrition would decrease peoples' susceptibility to health problems (e.g., non-communicable diseases such as cardiovascular attacks, diabetes, obesity) compared with health risks from diets that may be filling or satisfying but lack nutritional balance that can prevent the incidence of such diseases.

8.3 Effects of Continued Climate Change on Food Production

Food security is at risk for many of the Earth's regions unless food production can ramp up to feed increasing populations even as climate changes affect agriculture cropping and food animal husbandry. As noted previously, this is and will be more of a problem for countries in Africa and high mountainous regions of Asia and South America. In turn, this will cause social problems for water and food secure nations if hunger and thirst drive a 'survival' immigration towards them. For example, Europe has experienced the immigration stress because of wars/conflicts in the Middle East and Africa and long duration extreme weather conditions (e.g., drought) in Africa. These factors and others (e.g., seeking economic opportunities) have pushed perhaps 65 million people displaced from their homes or from their countries to migrate internally or emigrate since the beginning of the twenty-first century. The social problems in receiving countries that arise from a rapid and insufficiently monitored immigration (e.g., in the EU) can cause socio-political problems that can give rise to economic problems.

8.3 Effects of Continued Climate Change on Food Production

Sources of the global food supply depend on the availability of clean water whether from rain or irrigation sources. As planet warming continues, questions arise as to the demand-supply/water-food limitations as earth temperatures reach 1.5 °C above pre-industrial levels by 2050 or earlier (a high probability) and approach 2 °C (a viable possibility). In addition to the effects of a rising earth temperature, two other important controlling factors in future food production that were previously cited are the effects of geographic changes in rainfall patterns and rainfall variability, and extreme weather events (e.g., droughts, heat waves, flooding). The paragraphs that follow draw much from recent IPCC reports to assess how these changes would affect global food and water availability, especially in vulnerable, highly populated regions of nations, or in nations themselves [1–4].

8.3.1 Projections for Future Food Production

The IPCC reports show that during period from 1970 to 2019, there have been recorded increases in the intensity frequency, and duration of climate extremes (e.g., drought, heat waves, floods) and weather extremes (e.g., hurricanes [typhoons, monsoons], tornados/cyclones). The fore mentioned increase in the Earth's mean temperature to 1.5 °C or higher over pre-industrial temperatures as we approach mid-century will hurt the capacity to feed all in the growing global populations. Disruption of food production will be the case from an increase in extreme climate events such as heavy precipitation in many locations (e.g., bringing with it the potential for flooding, submergence of crops, soil erosion) and in others long lasting severe droughts (e.g., hurts agriculture production) [3, 4]. The strong population growth predicted for Africa and Asia and to a lesser degree in South America and the search for employment opportunities and education for children are driving grand migrations as rural citizens move to already highly populated urban centers (4.3 billion in 2020) to house 6.9 billion by 2050. Many of these cities are ill prepared to meet the challenges that global warming and climate change may bring to the food systems serving them. Governments now have to institute adaptations or modify those in place to improve agricultural production (crops and food animals), food storage, and transportation phases of their food systems that strengthen food security for their citizens.

8.3.1.1 Decreased Food Production by 2100

Computer models analyzed by IPCC experts and other researchers indicate that in some **low latitude regions where food production of staples is normally high (tropics and sub-tropics)**, a continued warming, precipitation variability (amount, intensity, and timing), and extreme weather events to 2050 and beyond, would make these regions more **vulnerable to declining crop yields and quality (e.g., for maize and wheat)**. It is mainly in low latitude countries that population growth to

2050 and beyond is predicted to be the highest putting their food security at risk. However, in many **high latitude regions** lesser effects of warming and favorable rainfall condition should result in better crop yields (e.g., maize, wheat, and sugar beets). In these regions, future populations will be stable or show slight increases in western Europe but will be contracting especially in eastern Europe to allow increased exports of their grains and other foodstuffs. Globally, however, the food production of staple crops is foreseen to decrease as the century progresses [3, 4]. The predicted decrease in staple crops as the century reaches 2050 and then on to 2100 is corroborated to a 'computer generated degree' by multiple studies a few of which are cited in the following paragraphs.

A 2012 meta-analysis of data from 52 original publications based on farm data from Africa and South Asia found that climate change has reduced yields of four of eight food crops studied and would continue doing so in the future. This was attributed to diminished rainfall and its timing with respect to dominantly rain fed crops and the global mean temperature rise. Across Africa, **average reduced yields** were found for wheat (−17%), sorghum (−15%), millet (−11%), and maize (−5%), whereas across South Asia, the estimated losses were of maize (−16%) and sorghum (−11%). No mean change was found for rice and insufficient data were generated for assessment of cassava, sugar cane, and yam crops. On the basis of their analyses, **researchers projected a mean change in yields for all crops of −8% in Africa and South Asia by mid-century**. As reiterated in previous chapters, these are regions with high estimated population increases to 2050 and beyond that put food security risk. **Proactive planning to increase agricultural production and plan implementation now is essential if future food security of these regions is to be stabilized and improved upon** [5].

A 2017 report focussed on how computer forecast **temperature increases** would reduce global yields of maize, wheat, rice, and soybean, staples that provide ~2/3 of populations' calories [6]. Table 8.1 shows the changes in yield of these crops projected for 2100 with respect to 1981 to 2010 average yields for IPCC RPCs (Representative Concentration Pathways) temperature limits. The 95% CIs (confidence intervals) for the RCP values are shown in parentheses.

The report further concludes that without CO_2 fertilization (no enhanced photosynthesis), without adaptation of improved farming techniques, and with no development or use of genetically modified crops, each 1 °C increase globally above the current mean temperature would result in an **average world loss in maize by 7.4%, wheat by 6%, rice by 3.2%, and soybean by 3.1%**. Nonetheless, there were instances where crops in some geographic zones showed gains in yields despite a 1 °C temperature rise [6].

The data in Table 8.2 illustrate the effect of increased global temperatures on how maize production is projected to decline for four major producers (not globally as in Table 8.1) if CO_2 emissions are not reined in and global warming were to reach 2 and 4 °C. Clearly, if maize production declines as forecast and global populations continue to increase, there is the question of whether there will be enough maize grown in the future to feed people and to provide feed/fodder for food animals (e.g., cattle, pigs). At these higher mean global temperatures, other food production

8.3 Effects of Continued Climate Change on Food Production

Table 8.1 Yield changes (%) caused by limiting temperature increases projected from the 1981–2010 means to 2100 for different scenarios of GHG emissions to the atmosphere. Slightly modified from [6]

Scenario	Maize	Wheat	Rice	Soybean	Mean
RCP2.6 at <2.0 °C	−8.6 (−18.6, −1.8)	−6.9 (−15.0, −1.4)	−3.3 (−9.2, 0.8)	−3.6 (−11.2, 1.7)	−5.6 (−14.4, −0.1)
RCP4.6 at <2.4 °C	−14.2 (−27.9, −4.9)	−11.4 (−21.7, −3.9)	−5.5 (−13.8, 1.0)	−5.9 (−17.0, 3.1)	−9.2 (−21.1, −0.3)
RCP6.0 at <3.0 °C	−17.4 (33.1, −5.8)	−14.0 (−25.7, −5.1)	−6.8 (−16.8, 1.3)	−7.2 (−20.2, 3.6)	−11.3 (−25.6, 0.1)
RCP8.5 at <4.9 °C	−27.8 (−50.4, −9.7)	−22.4 (−40.2, −8.5)	−10.8 (−25.3, 2.4)	−11.6 (−31.0, −6.0)	−18.2 (−38.6, −0.7)

Table 8.2 Projected global warming changes in maize yields for 2 and 4 °C compared with the average of the 2012–2017 yield for the top four producing countries that export 87% of the world's maize. Statistical spreads are in parentheses [7]

Country/rank	% Change with 2 °C warming	% Change with 4 °C warming
1. United States	−18% (17.4–18.3%)	−46% (45.4–47.5%)
2. China	−10% (10.1–10.7%)	−27% (26.7–28.0%)
3. Brazil	−8% (7.6–8.1%)	−19% (19.0–19.9%)
4. Argentina	−12% (11.3–11.9%)	−29% (27.9–29.0%)

sources will suffer as well. **Concurrent crop losses** >10% from the major producers listed in Table 8.2 would mean famine and starvation for millions of citizens but a calculated probability of such a simultaneous crop loss at the current warming level (1 °C) is basically zero. However, at a 2 °C global warming rise, this probability becomes 7%, and with a 4 °C rise the probability soars to 86%. As already suggested in the lead paragraph of this section, there is ongoing research into the breeding of heat tolerant grain species for optimum yield (production) and quality of food crops (e.g., maize, rice, and wheat) at higher temperatures. Variability in precipitation and corresponding changes in soil moisture also affect maize crop yields but this is outweighed by the warming to 2 °C or higher [7].

In addition to cereals, vegetables and legumes are essential to healthy diets. A review was made of 174 journal articles published between 1975 and 2016 on effects of environmental change on vegetables and legumes in 40 sub-tropical countries (in southern Europe, North America, and southern Asia) [8]. For research articles using a 20 °C baseline temperature, the consensus was that a 4 °C rise in temperature could **reduce mean yields** by −31.5% (95% CL −41.4% to −25.5%). At temperatures <20 °C, the mean change was +34.9% (95% CL −47.9 to +117.6). The different papers also considered one or more than one impacting factor in addition to temperature (e.g., CO_2, O_3, water availability, and salinity). However, the methodologies followed in the 174 studies were not consistent geographically or in their modus operandi so that the results cited should only be taken as suggestive of the general expected impacts.

Overall, the general consensus of climate scientists and their agronomic experts is that climate changes will cause an aggregate reduction in future agricultural productivity especially in low latitude regions. For major crops in **the tropics and temperate regions** (wheat, rice, maize), the IPCC reports that a local temperature increase to 2 °C or greater above the pre-industrial baseline will reduce food production and availability in the **Sahel, southern Africa, the Mediterranean, central Europe, and the Amazon** compared to the 2020 food production at the 1⁺ °C temperature increase above the pre-industrial baseline temperature [2]. To this end, Table 8.1 shows estimates of the effects of temperature rise on major crop yields by 2100 if adaptations are applied to keep temperature rise to <2 °C (RCP2.6: reduce and stabilize CO_2) and the effects if little is done to stem a global warming that could reach <4.9 °C (RCP8.5) as the century progresses. In addition to crop declines, livestock growth and production will suffer because of impact on water availability, feed (fodder) yield and quality loss, and disease. Most living natural ecosystem resources will be adversely affected by earth temperature increases whether in the range of 1.5–2 °C or higher.

Discussions about regional agricultural production responses to climate change have to be tempered by the fact that there can be a great variability between sub-regions. As observed globally, pockets of farmland can have good agricultural yield within large geographic regions that suffer aggregate losses of food. This is attributed to favorable micro-climate conditions affected by topography, or proximity to large water bodies, or the crops sown and adaptation(s) utilized. For example, a 2018 study predicted better yields of maize, rice, wheat and possibly other cereal crops in some tropical/subtropical areas of **Sub-Saharan Africa, SE Asia, and Central and South America** where crop loss was expected [2]. Adaptations to counter the losses include seeds used (hybrid or GMO) that are drought and/or heat resistant, disease resistant, resistant to herbicides used to control weeds and pests, and ability of a crop to recover after submergence. Another important adaptation has been the efficient use of available water supply via drip irrigation or other water conserving efforts during drought for the 86% of rain-fed farmland. If water is available, irrigating fields during the non-growing season can be beneficial to later cropping. As reported in earlier paragraphs, expert consensus indicates that if not reined in, climate changes in future decades will see diminished global food production.

8.3.1.2 A Potential Source for Increased Food Production

There are many regions worldwide where crop yields are limited because of saline soil. About 20% of irrigated cultivated agricultural fields (~45 million ha) have saline soils at different salt concentrations [9]. These are mainly in warmer temperature arid and semi-arid regions. Salt content of a soil inhibits a crop's capability to take up water from the soil. This affects food production by interfering with the uptake of the nutrient nitrogen by plant roots. A restricted nitrogen uptake reduces growth (crop yield) and can interfere with plant reproduction. Each crop whether grain, grass, vegetable, or fruit has a different tolerance to a saline level (Table 8.3).

8.3 Effects of Continued Climate Change on Food Production

Table 8.3 Soil salinity levels and response of crops. The levels are given as ranges of electrical conductivity (EC) measurements that relates to soil salinity. EC will vary with soil texture such as sand, loam, and clay, and mixtures thereof. One dS/m (decisiemens/meter) can be taken as 600 ppm (0.06%). Seawater salinity is 35,000 ppm or 3.5% [9]

Generalized salinity level	EC (dS/m)	Effects on crops
Non-saline	0–2	None to negligible
Slightly saline	2–4	Yields of sensitive crops may be limited
Moderately saline	4–8	Moderate to severe—yields of many crops restricted
Strongly saline	8–16	Severe—only tolerant crops have adequate yields
Very strongly saline	>16	Very severe—only very saline tolerant plants have adequate yields

An adaptation currently used to cultivate crops in saline soils is to identify naturally saline tolerant species and use their seeds. An extensive list of plant species and their saline soil tolerance limits has been compiled by the FAO [10].

One group, Salicorp, approached the problem of making seeds tolerant of soil salinity by **reportedly** developing a wet chemistry technique over a few years period of testing. Seeds for a given crop are soaked in a chemical solution (proprietary) under strict laboratory conditions. Seeds for different crops are soaked in cocktails tailored specifically to each and to the measured soil salinity. The group announced that trial plots with treated seeds grown side by side in fields with untreated seeds had a 12–30% higher average yield in representative high and extreme saline soils and semi-arid environments in India (rice), Israel (sweet corn and silage wheat), and other crops in Spain. The so-called **'proof-of-the-pudding'** will come with the crop yields from commercial scale field trials that are **reported** to be underway for rice, wheat, capsicum (peppers), and tomatoes (Pers. Comm., Salicorp. Sept. 10, 2020). A potential benefit if the method proves commercially viable is that fields with treated seeds could be irrigated with brackish water.

8.3.2 Fisheries

The Earth's human carrying capacity with respect to food is supported by availability of food fish. The FAO reports that in 2016, of the 171.6 million tonnes of fish harvested globally, people consumed 88%. The other 12% went for non-fish use (e.g., fish meal, fish oil, fertilizer). The wild fish catch has leveled off in recent years in accordance with international or local marine protected areas that aim to prevent overfishing and allow for procreation and reestablishment of ocean fisheries that otherwise were being overfished. Aquaculture production (freshwater, brackish water, seawater systems) has been increasing globally and in 2016 supplied 53% of the total of fish caught for food when fish meal or fish oil are not considered, or 47%

otherwise [11]. World **aquaculture** production is highest in Asia (especially in China and India) with 89.4% of the total followed by the Americas with 4.2%, Europe with 3.7%, and Africa with 2.5%.

Fish are an important protein-rich food source. They supply 17% of the global population annual animal protein intake. Food fish are especially important in Asia where ~3.2 billion people in urban centers and in coastal villages and towns get ~20% of their protein from fish consumption. That food fish harvesting is increasingly important in food security planning is emphasized by the fact that from 1961 to 2016, the average annual increase in fish consumption worldwide was 3.2% or twice the average rate of population increase. It rose from 9 kg/per capita in 1961 to 20.3 kg per capita in 2016. The 3.2% annual consumption increase for the 1961–2016 time frame rate was greater than the 2.8% annual consumption increase for meat from all land animals [11]. Wild fish harvesting and fish from aquaculture are now threatened by effects of global warming/climate and will likely be at greater risk in the future. Adaptation is the key to sustain and improve fish production from these sources while minimizing environmental impact on the ecosystems that sustain them.

It must be noted that marine fisheries suffer from rogue trawlers that illegally harvest fish in territorial and international waters without care as to limit, thus depleting stocks for reproduction and a sustained yield. When identified, these vessels are seized and impounded for later sale with fines for owners of the ships and in some cases identified buyers of illegal catches.

8.3.3 Warming Temperature Effects on Food Fish

Responses of fish and shellfish in open sea and aquaculture fresh water, brackish water, and seawater fisheries to global warming/climate change affects the food fish contribution to the Earth's sustainable food supply. One response is the poleward migration of ocean food fish or prey that predator food fish follow from warming habitats that disrupt their food supply and procreation to more welcoming cooler waters. This may be either in the upper 75 m (42 fathoms) that have warmed by 0.44 °C (0.79 °F) from 1971 to 2010 or from a lesser degree of warming at water depths to 700 m (393 fathoms) [1]. A second response is for cultivated fish that do not grow as well or as quickly in warmer waters in aquaculture ponds and brackish or sea water pens. Because of poleward migration of some food fish/shellfish, fishing fleets have to sail to different fishing grounds where there are enough fish to harvest while leaving enough to procreate for future catches. For aquaculture, it is necessary to adapt to grow species that naturally tolerate the warmer temperature conditions in their enclosed habitats or to cultivate genetically modified species.

8.4 Other Warming Threats to the Food Fish Supply/Security

Another threat to the food fish supply and ecosystems where fish flourish is the warming at coral reefs. Coral reefs at shallow sunlit, warm, alkaline waters and low light coral reefs at depths of 40–150 m are habitats for species important to fisheries. These and deeper (to 2000 m) cold water reefs are found in <1% of the ocean. The sunlit and low light coral reefs support one quarter of all marine life by providing more than 4000 species of fish, with food, shelter, and spawning sites. As seawater temperature rises at coral reefs and there is too much exposure to sun for some reefs, symbiotic zooxanthellae that coral polyps depend on for food die. The result is that coral polyps lose color and they die leaving their calcium carbonate exoskeletons that can not support marine animals that lived, fed, and spawned new life at healthy coral reef ecosystems. Australian scientists are working on transplanting coral polyps from the warming (bleaching) reef ecosystem of South and Southeast Australia to more hospitable environments in the Great Barrier Reef ecosystem.

The increasing concentration of CO_2 in the atmosphere and hence its partial pressure increases its concentration in sea water alters the acid/base condition or pH of sea water by making it less alkaline. This has been called acidification. The pH measures values between 0 and 14. A value of 7 represents a neutral water, neither acidic nor alkaline (also referred to as basic). A pH <7 is acidic and a value >7 is alkaline. The 'normal' pH of seawater is ~8.1 (can range from 8.08 to 8.33). Less alkaline (but not truly acidified) ocean pH conditions negatively impact marine life and disrupt the oceanic food web by diminishing the ease and fullness of shell growth and reproduction of zooplankton, tiny crustaceans that many fish feed on. In less alkaline seawater, it becomes harder for corals to build their calcium carbonate ($CaCO_3$) skeletons (reefs) and serve as spawning grounds for many food fish. The same is true for shellfish (e.g., oysters, clams, mussels) and other large and small shelled forms as they work to build their shells and keep their place in the marine food web. This threatens the 'normal' continuity of the ocean food web and can reduce the world's wild capture food fish/shellfish supply [12]. A specific example is salmon in the wild that as fry (new hatched salmon) initially feed on zooplankton, the tiny crustaceans that comprise the base of the ocean food web. Acidification makes it difficult for the crustaceans to form their shells fully and this affects their life cycle and the salmon fry food supply of krill, crustaceans 1/4 to 2 in. in size, and small fish. With enough food to sustain normal growth, the fry mature to adult salmon that are captured and are a good food/protein/fat source for human consumption.

Unlike CO_2 for which partial pressure increases as more is emitted into the atmosphere and more is dissolved in ocean water even as it warms, the atmosphere's O_2 partial pressure is stable. Nonetheless, there are some oxygen minimum zones (eutropic waters) that develop from sea water density stratification or masses of decaying organisms or dumped urban wastes. Fish avoid these zones if they can because they would otherwise struggle to pass enough sea water through their gills to fulfill their O_2 needs. The 2019 IPCC report on the changing climate and effects of melting ice sheets and glaciers on the oceans and their ecosystems discusses in fine detail the effects on fisheries that were cited in earlier paragraphs (warming waters, migration of ocean species, coral reefs as spawning sites, acidification, and oxygen minimum conditions) [4].

Projections as to ocean fishery and aquaculture contributions to the food carrying capacity (food availability) or lack thereof in the future if CO_2 emission to the atmosphere increases at its present rate or at an increased rate indicate that there will be **increased food fish productivity at high latitudes** and **decreased productivity at low and middle latitudes** (see Fig. 7.2). This reflects the global warming caused poleward migration of food fish and their prey and a 10% mean decrease in fish/shellfish populations from tropical and subtropical waters. Some computer models suggest that there will be a **marine fishing loss** of 1.5 million tonnes (3.3 million kg [72.6 million lb]) if the Earth warms by 1.5 °C and a >3 million tonnes (>6.6 million kg [145.2 million lb]) loss at a 2 °C warming [2]. As suggested earlier, this protein laden food source could be lost to coastal villages/towns populations that need food fish to sustain healthy lives and in some cases their economies. It would be a loss as well to growing urban populations that include food fish in their diets and cannot afford to be deprived of this protein source. A hope is that aquaculture production will make up for the loss from marine fisheries as ocean fish stocks that have been overfished are given time to recover before controlled harvesting can begin again.

8.5 Hypothetical Solution to Feed Growing Earth Populations

8.5.1 Diet Change = More Land for Food Production

A 2016 paper evaluated the relation between land needed for food production and a nation's average diet. Researchers selected three of what **they considered as representative** average diets for determining future global agricultural land use if consumers worldwide converted to one or another. The first, an average diet is followed by ~30% of India's population and would result in **less acreage used for food production** than existing global levels. A second, the average diet followed by Mexico's population, would necessitate additional land for cropping and animal husbandry but would be feasible globally within reasonable limits of conversion of natural ecosystems to agricultural land. The third, an average diet followed by the United States population, would not be feasible worldwide even if all habitable land were converted to agriculture. In this latter case, about 11.2% of the global population (in 19 countries) follow a diet heavy in animal source foods that require vast tracts of land for feed cropping and grazing for food animals. In theory, if the world population converted to an **'average' Indian** (vegetarian) **diet**, the researchers estimated that **55% less land** (feasible) would be needed for food production. Conversely, a global shift to an average United States diet would require **178% more land** (not feasible) to feed the world population [13]. As the Earth's population continues to grow, these theoretical numbers will change but the proposition that a global dietary revolution to a **plant-based diet** from an animal-sourced diet would best serve the health and welfare of the world population remains a valid concept. This is the subject of the following discussion.

8.5.2 Healthy Diet for Ten Billion People in 2050?

The Earth's capacity to provide food for a maximum population of ten billion people predicated on a plant-based diet had been proposed in 2002 [14] and served as a foundation for a later detailed and extensive 2019 report by an EAT-Lancet Commission. The report considered population growth, and posits a solution to the nutritionally sufficient sustainable feeding of a ten billion person 2050 population that would also contribute to global physical, biological, and mental health. The basis would be a phased shift during the 2020 to 2050 time frame to a diet of mainly plant-based foods (e.g., **followed by only ~30% of India's 1.4 billion population**) from existing diets of animal-sourced foods for many populations. The report presents a healthy reference diet from which consumers can choose that provides ~2500 calories per day from ingestion of ~1.3 kg/day of mass. This diet is dominated weight wise by whole grains (232 g), vegetables (colored and starchy 300 g), fruits (200 g), and dairy foods (250 g), with lesser but dietary important amounts of protein sources (84 g), legumes (100 g), tree nuts (25 g), added fats (52 g), and added sugars (31 g). The diet allows for intake changes within preferred ranges that reflect a population's preferences, cultural norms, and location with respect to a food source. The Commission also estimates that a global change to healthy diets would save about 11 million lives annually [15].

The question is whether enough consumers of animal-sourced foods (natural or processed), saturated fats, sugars, and high calorie comestibles can be weaned away from them to healthy plant-based foods. The same question would apply to consumers entering the middle class in many countries with incomes that give them access to less healthy diets. If large numbers in animal-sourced diet populations made a shift away from them, what might this mean for food production and food security? It is important to know that in 2020, ~38% of the Earth's ice-free land is devoted to agriculture (cropland and grazing) and that **on the cropland, 2/3 of all soybean, maize, and barley crops and 1/3 of all grain crops are grown for animal fodder (e.g., for cattle [beef, dairy], pigs, chicken)**. It is also important to know that to satisfy an increased population demand for unhealthy dietary food and the profits they bring to providers, natural ecosystems have been invaded and razed with a loss of biodiversity necessary for preserving natural resources for human use. For example, this invasion has happened and is happening in Brazil, the Democratic Republic of Congo, and areas in Southeast Asia.

Hypothetically, with an educational and a health driven global population transition towards a mainly plant-based diet, the need for animal feed would drop dramatically and newly available cropland can be repurposed to grow plant-based comestibles for human consumption without invading and altering natural ecosystems for agricultural exploitation [15]. For example, in the United States alone, 30 million ha are cultivated for livestock feed grains, grains that could otherwise feed 800 million people. With less livestock production worldwide, more grain production could help sustain food needs for increases in population. A like change as electric and hybrid vehicles take over the automotive market would repurpose farmland used for biofuel crops (e.g., maize, sugar cane) to grow crops for people. Climate changes in the future will interfere with agriculture as changes in precipitation patterns affects some or all of today's 80% of rain fed cultivation that provide 60% of food crops and 20% that are irrigated with water from lakes, rivers, and aquifers and provide 40% of food crops. Regions with great dependency on rain fed cultivation and some with high population growth include Sub-Saharan Africa (95%), South America (90%), Mid East and North Africa (75%), East Asia (65%) and South Asia (60%). There are questions of where and to what extent the locations of warming and precipitation changes will affect food production/food security as the earth warms from the 2020 1^+ °C above a pre-industrial temperature to what will likely be 1.5–2 °C rise by mid-century.

8.5.3 In Theory, in Practice

In theory, according to the EAT-Lancet Commission's estimates, **it would be possible feed ten billion people** by 2050 with healthy plant-based diets from sustainable food production systems. This would be the case because land used for animal food could be repurposed for plant-based foods as noted earlier for grazing land no longer necessary for food animal feeding. However, the change **would be unlikely to be able to feed a larger population (e.g., 11+ billion in 2100)** [15]. An implication here is that the food system **without** a shift to plant-based diets could not feed the projected 2050 population especially with food production threatened by effects of increasing global warming (see Tables 8.1 and 8.2). Would a realistic assessment on a limit of food production as it stands now be able to feed **8.5 billion, 9 billion** people enough of a nutritious diet if the expectation is to bring down and eventually eliminate global hunger?

In practice, a 'Great Food Transformation' from animal-sourced to plant-based foods, the result of a stimulating thought provoking exercise, is unlikely to take place given the reality of numbers. The number of vegetarians in the world is hard to come by but has been estimated from country census data. Globally, India has the highest number of vegetarians at **420 million** of its 2020 population. The total number of vegetarians listed for the world by Wikipedia was 620 million people (out of

7.8 billion people [8%]) [16]. In the Wikipedia listing **375 million** vegetarians are attributed to India. Given these numbers, it is unlikely that a population committed to a plant-based diet in the decades leading to 2050 and beyond, increase as it may, will significantly influence food production/food security for a possible ~9.9 billion people (if that number is reached?) that could have a larger number of consumers of animal-sourced foods than in 2020.

8.5.4 Diet Boost from Kitchen Prepared Insects?

It should be noted that insects can be a food source of protein, fats, and minerals. In some countries in Asia, Africa, and the Middle East (e.g., Thailand, DR Congo, Israel), insects are prepared and eaten as snacks. Fried or roasted short horned grasshoppers (locusts) are favored snacks rich in protein (50% dry weight) but not important to daily diets. Given the swarms of hundreds of millions of desert locust that attack crops in East Africa annually, the question might be asked as to whether they can be captured in sufficient numbers so as to add significantly to a food supply. As noted above, they can be an important source of protein, fats, and nutrients where/when other food sources are in short supply. Demonstrably, insect consumption can be an important food source. In Ghana, when food is scarce during springtime and planting is in progress, rural Ghanians gather termites that come out of their nests during spring rains and eat them, roasted or fried, as their main source of protein and fats.

8.6 Some Foreseen Effects of a Rising Global Temperature

As reiterated previously, during the early decades of the twenty-first century, the Earth's mean temperature reached 1 °C above the pre-industrial baseline. Earlier chapters discussed how this level of warming contributes to the shortfalls in food production and water deficits in different regions. The warming continues. Some researchers estimate that world governments have a two to three decade window to work together to stop the increase of CO_2 into the atmosphere in an international effort to keep global warming to a 1.5–2 °C range above the pre-industrial temperature to slow and stabilize climate conditions. Holding this temperature line could reduce future warming damage to rain fed or irrigation dependent food production and also reduce the number of people exposed to water stress by up to 50% [2]. Unfortunately for people and for ecosystems that yield natural resources that support humanity, this may not be achieved for reasons discussed in following paragraphs.

8.6.1 Reject Coal or Not in Industrial Plants or as Export?

Given political and economic aims of some countries to support economic (industrial) development in spite of its damaging impacts on the planet and its populations, it is highly unlikely that the Earth's mean temperature will stabilize at 1.5 °C. Coal-fired power plants continue to function and increase in number, especially in China that in 2017 world's leading emitter with ~28% of CO_2 emissions followed by the United States with ~16% and India with 6.1%. Coal-fired power plants in China generated 980 GW (gigawatts) of electricity in 2016. China is building or plans to build ~200 coal-fired plants that will raise the electricity capacity 250 GW (~25% increase) with additional emission of CO_2 to the atmosphere. China is also abetting (financially and with Chinese technicians and workers) the construction of many coal-fired power plants in less developed and developing countries in Africa and Southeast Asia although power sources other than coal-fired power generating options may be available (hydropower, solar, wind, natural gas, geothermal). Coal fired power for development and the financial benefits it may bring clouds a realistic vision of harm to public health and to environments/ecosystems from emissions of CO_2, fine particulates, and toxic metals.

There are **economic realities** with respect to power generation that can benefit the environment. The principal reality is long-term availability of fuels cheaper than coal for power generation. Each country accepting the construction of coal-fired power generating facilities is not doing its due diligence if it does not evaluate the long-term costs of other than the coal option. For example, the availability of cheaper natural gas in the United States has greatly reduced the use of coal in power plants and other industrial sectors resulting in much less emission of CO_2, heavy metal toxicants, and fine size particulates. This fuel shift in the United States mainly to natural gas has taken place even though environmental regulations vis-a-vis the use of coal have been greatly reduced by the 2016–2020 Trump administration in order to support coal companies, but to no avail as many major coal mining operations have shut down. During 2020, global electricity generation by renewable resources overtook fossil fuels in general in the EU countries, led by wind and solar as EU nations strived to reduce CO_2 emissions and adhere to the Paris Agreement.

8.6.1.1 Coal: Export or Not/Develop Mines or Not

Australia has been the major exporter of cheap coal to the Asian coal-fired power plants and has been under much pressure to stop supplying the commodity. However, coal has brought in 15% of the country's export income and is second financially only to the export of iron ore and concentrates. Because of this the government self-interest mode is not yet ready to lose the income that is important to its GDP and lose a source of employment that would hurt many Australians. During the fourth quarter of 2020 Chinese put restrictions on the import of Australian coal because of Australia's non-support of China's territorial disputes and acceptance of immigrants from Hong Kong. In the meantime, China will import coal from Mongolia, Indonesia, and Russia. China is already operating coal mines in Zambia and Zimbabwe and in 2019 had reached an agreement with the Zimbabwe government to begin environmental assessment studies at two locations inside the Hwange national park previous to geological/geochemical surveys, exploration drilling, and land clearance for road building. If exploration were to begin and lead to exploitation, habitats of endangered species such as the black rhino, pangolin, and tinted dogs would be disrupted along with those of elephants, buffalo, giraffes, and cheetahs. Income to the tune of hundreds of millions of dollars from safari tours for the country and locals could be lost [17]. This could have been an example of a less developed country's push to short-sighted gains at the expense of long-term loss as coal mining takes place with the pollution it generates and ultimately ceases when the coal is played out. However, in 2020, legal and political/economic pressure from the Zimbabwe Environmental Lawyers Association, international organizations, government agencies, and NGOs protective of humans and wildlife environmental health 'convinced' Zimbabwean government leaders to review and reverse their agreement and stop Chinese coal company programs, thus preserving the integrity of the Hwange and all other national parks.

China has 96 billion tons of coal reserves to drawn on in addition to the coal it imports. This may explain in part why China is building many new coal-fired plants with many others in stages of planning. In no way is this going to help reduce CO_2 emissions into the atmosphere in the future and certainly is counter to the Chinese President Xi's pledge to get to zero emissions by 2050/2060. In addition, China is financing the construction of many new coal-fired power plants in South Africa, Pakistan, Bangladesh, Indonesia, Viet Nam and other countries. This defeats the essence of the country's pledge to help reduce global CO_2 emissions in the effort to slow and stop global warming. Only if China and other nations that generate most of their energy needs by combusting fossil fuels such as Indonesia and South Africa collaborate and are compliant in **now** reducing their reliance on fossil fuels can the attack on global warming achieve success in slowing the warming rate and then stopping the warming. This would be the good for all globally, not self serving that ultimately would bring disaster to all.

8.6.2 Forecast of Regional Temperature Effects

IPCC and other research reports are in agreement with computer analyses and interpreted predictions that there will be temperatures higher than 1.5 °C in many regions if the Earth's mean temperature reaches 1.5 °C or higher. In the mid-latitudes, the number of extreme hot days is projected to increase by up to 3% at a 1.5 °C rise and up to 4% with a 2 °C rise. In both scenarios the hottest days are expected be in the tropics. High latitude locations are predicted to seasonally suffer nights of extreme cold [2]. As earlier noted, rises in temperatures are expected to cause regional increases in the amount and intensity of heavy precipitation while other areas will suffer from increases in intensity, frequency, and duration of heat waves and drought (e.g., Mediterranean region and Southern Africa). Such heat wave and drought conditions during 2019 and 2020 dried out vegetation and created ideal conditions for lightning strikes ignited severe and extensive wildfires at different locations worldwide (e.g., South and Southeast Australia and Western United States, Western Canada, Alaska, Siberia, Brazil, Lebanon, Syria, Iraq, Israel, Angola, and Portugal). The wildfires have killed people, destroyed hundreds of structures, caused evacuations of hundreds thousands of people, and poisoned the atmosphere locally and regionally with fine size particulates and other toxicants. During the course of the century, such extreme weather conditions and the disasters they cause will likely diminish the planet's food carrying capacity regionally in increasingly warmer areas thus disrupting internal food systems and food exports to countries depending on them.

It is intuitive that if (when) the Earth's mean temperature reaches higher than 1.5 °C, probably to 2 °C or more above the pre-industrial era baseline, the risks to people from extreme weather disasters such as mentioned in the previous paragraph will increase. Food production will suffer in such areas depending on the duration of a climate event. The IPCC cites several locations that would be at increased vulnerability from a 2 °C rise. These include **high latitude and/or high elevation northern hemisphere** regions in countries of eastern Asia (e.g., China, North Korea, South Korea, Japan, Taiwan), and eastern North America (Canada, United States). A rise to 2 °C from the 1.5 °C level will accentuate the harm to terrestrial, lacustrine, and coastal/estuarine ecosystems and the loss of biodiversity. Agronomists evaluated computer generated data and forecast that without adaptations in food production methods to cope with higher temperatures, crop yields in many regions will lessen even as growing seasons may be longer [2].

It is important to note a non-food related global warming ramification of a gradual rise in Earth temperature to 2 °C in parts of the **higher latitudes**, (e.g., Siberia, Alaska, Northern Canada, Scandinavia). This is an increased thawing of permafrost. Depending on the amount of temperature rise, the thawing could release large volumes of the GHGs methane, carbon dioxide, and nitrous oxide into the atmosphere thus adding fuel to global warming. This presents a problem that can be damped only by stopping a temperature rise.

For temperate, sub-tropical, and tropical climatic zones with temperatures that gradually increase to 2 °C, the production of the life sustaining crops wheat, rice, and maize are predicted to decline in yields and quality. An adaptation that may counter this threat is the earlier discussed hybridization of plants (natural and natural-assisted) or plants that have been modified by genetic manipulation to produce seeds to grow plants that tolerate higher temperatures. If this plant resistance to a temperature rise to 2 °C is not achieved, researchers estimate that there **will be reduced food availability** in southern Africa (with growing populations) and the Sahel (see following paragraph), and in the Mediterranean region, central Europe, and the Amazon [2].

The Sahel is of unique interest because of its geographic reach and the 2019 populations of the countries (~336 million [206 million in Nigeria]) that contain part of the Sahel. More than 150 million citizens in those nations northern regions depend on Sahel agriculture for food crops and grazing for herded food animals. Geographically, the Sahel is a semi-arid region that presents a transition zone between the arid Sahara desert to its north and an arcuate multi-countries stretch of humid savannas to its south. The Sahel western end is in Senegal at the Atlantic Ocean end and extends eastward through areas of Mauritania, Mali, Niger, Nigeria, Chad, Sudan, to Eritrea at the Red Sea, a distance of more than 3800 km (2360 mi). Its major rainfall of 100–200 mm (4–8″) is during June, July, and August. A long duration drought in late 1968–1974 caused famine in the region with deaths in the 100,000s, loss of cattle herds, and desertification as the Sahara desert extended scores of kilometers into the Sahel [3]. Such drought events will likely occur in the Sahel as the twenty-first century and global warming progress and threaten the food supply for citizens in the northern reaches of eight countries. An ominous portent for the future is a 2020 drought that reduced the Nigerian rice production by 25% stressing that country's food security. The 2050 population of countries with part of their populations in the Sahel is estimated to more than double by 2050 (~684 million [~401 million in Nigeria]) [18]. The number in the Sahel will likely show significant increases as well. To counter the advance of desertification resulting from drought and protect agricultural land, the Sahel nations supported by Sub-Saharan nations started the planting of the Great Green Wall. This is a 30 km wide barrier of drought resistant native vegetation (e.g., Acacia trees) to stop the advance of the Sahara desert and land degradation/desertification. The plantings are almost 20% done with 20 million ha planted by 350,000 workers at an investment from 2007 to 2018 of US$90 million dollars.

8.7 To Slow, to Stop, to Reverse Global Temperature Rise

Again, we can speculate on how to slow and stop the warming that threatens food production and perhaps ultimately initiate a controlled cooling. An achievable first step is to greatly reduce the emission of CO_2 and other GHGs into the atmosphere from industrial sites (e.g., coal-fired power plants, cement plants, and many others) by installing technologies that can chemically scrub up to 80–90% of CO_2 out of flue gas. This can be done by absorbing CO_2 into a medium or adsorbing the gas onto a medium, for later desorption and recovery to generate CO_2 to use in oil fields for secondary recovery (see Sect. 4.7, second paragraph), to precipitate salable minerals or loop to make pure CO_2 to sell. Such systems were operational globally during 2019 at 17 industrial facilities and prevented the emission of more than 31.5 million metric tons of CO_2 into the atmosphere, of which 3.7 million metric tons are stored in subsurface porous, permeable geologic formations that are bounded above and below by impermeable rock to prevent leakage [2, 19]. In addition to operational facilities, there are 65 commercial carbon capture storage (sequestration) facilities (CCS) in development globally. This requires investment in scrubbing systems and their management to make sure that equipment is installed, used, and maintained. A hub-centered shared infrastructure for transport and storage sites for the captured gas would encourage industries adoption of CCS systems. During 2020, flue gas captured mass of CO_2 was reported as being 40 million metric tons. To perhaps stem global warming (zero emissions) this would have to increase to 140-fold to 5635 million metric tons annually by 2050 [20]. One would hope that nations agreeing to adhere to the Paris Agreement CO_2 reduction levels make investments as necessary and set rules for equipment use and maintenance schedules. Captured CO_2 has also been used for the commercial production of important solid compounds such as calcium carbonate, magnesium carbonate, calcium bicarbonate, and others that are used in various products. Recently, captured CO_2 has been made available to be piped into greenhouses to stimulate photosynthesis and plant growth, and is being used in the commercial preparation of carbonated drinks such as colas.

An achievable second step to increase the capture of CO_2 from the atmosphere is the reforestation of timbered forests, forests cleared for farming and animal husbandry (e.g., in the Brazilian Amazon), or wildfire razed forests worldwide. Afforestation will add to the Earth's CO_2 sink capacity. Ocean sink capacity can be increased by stimulating growth of algae, other sea plants, and plankton but only to the degree that is not harmful to other sea life as would be a 'red tide'.

In contrast to the increasing industrial emissions to the atmosphere, agriculture's contributions of GHGs (methane CH_4, nitrous oxide N_2O, and CO_2) released to the atmosphere from crops and livestock, and from land use (tilling) are in decline. There has been a steady decrease in the contribution of agricultural GHG releases to the atmosphere from 29% during the 1990s decade to 25% in the 2000s decade to 20% in 2017, with 11% of the 20% from livestock (flatulent activity) and crops, and 9% from land use.

8.7.1 Direct Extraction of CO_2 from Air

A rather recent approach being evaluated for CO_2 removal from the atmosphere is direct air capture (DAC) of CO_2 by sucking air into a system where the captured CO_2 is sorbed onto an active chemical surface or into a chemical medium for later desorption and use or storage or the CO_2 is reacted with a chemical to form a solid compound. The remaining air is released back into the atmosphere. Carbon Engineering based in Canada has successfully tested its system and has paired with investors (1Point Five) and Occidental Petroleum Company CCS expertise to build a plant during 2024 in the Permian Basin, West Texas, USA that is planned to capture 500,000 or more metric tons of CO_2 annually at an estimated cost of $100 per metric ton. Another company in Switzerland has a plant in operation that is capturing 900 metric tons annually but at a prohibitive estimated cost per metric ton [21, 22]. Global Thermostat is building a plant in Alabama, USA with Exxon-Mobil that is commercially purposed to sell CO_2 to carbonate beverage companies and provide raw material for plastics manufacture. There is a Future Act in the United States Congress that if passed would provide this company with a $35 per ton CO_2 subsidy

[23]. Cost estimates for the different ventures vary but realistic company figures based on demonstration plant data suggest that initially removal costs would be in the range of $130 ($90–$150) per ton CO_2. The expectation is that this price will be reduced as many plants are constructed with efficiency improved through chemical engineering such as by using renewable energy for electricity thus reducing heat and cutting fossil fuel emissions. This would be similar to price reductions achieved for solar and wind generated electricity. A 2019 academic assessment suggests that technology could reduce the price of direct air capture plants from a 2020 cost estimate of €133 per ton CO_2 captured to a 2040 estimate of €40 per ton CO_2 captured [24].

Given that 37 billion tonnes of CO_2 were added to the atmosphere in 2019 [25], individual plants would hardly affect the CO_2 content of the atmosphere. If the 500,000 tonnes facility cited above performs as planned, the world would have to have 748 like plants to reduce the CO_2 content in the atmosphere by 1% complemented by the mass of CO_2 taken up by the sinks and the CO_2 captured at emission sources. At a conservative cost $130 per ton CO_2 extracted instead of the estimated $100 per ton cited above, the annual cost for 748 plants would be ~$55 billion or for a decade ~half a trillion dollars. This is readily achievable if international banks make low cost/no cost loans to less economically advantaged nations and richer nations contribute to the construction, operation, and maintenance of the plants. What is possible is that **a four pronged attack (capture CO_2 at source or reduce gasoline use [industries and vehicles], improve sinks, reduce agriculture output, and direct air capture of CO_2) if put into action, can slow global warming of the planet and ultimately stop it at a point where the planet is livable for all citizens**. However, this has to be a committed international effort to meet the Paris Agreement goals to slow and stop global warming. It means that nations using coal-fired plants for electricity in their development efforts either switch to other sources to generate energy or use and maintain CO_2 capture technology with ability to utilize or dispose of CO_2 securely, as needed (see Sect. 8.6). **The time to strike with such approaches is now in order to sustain and improve the Earth's human carrying capacity in the 2020 decade and subsequent three decades even as populations grow. Delay will bring on further negative impacts on societies worldwide as populations grow, global warming continues, and climates change**. If China does not modify its latest 5 year plan and reduce construction of coal-fired plants nationally and its support for financing coal-fired plant construction in several other countries, the chances for slowing and stopping global warming are minimized. Only economic pressures from signees to the Paris Agreement on China can bring its coal-fired power plant development to a halt and support the worldwide effort bring global warming under control.

If we assume that investment in direct air capture is the option to follow with hundreds of high mass capture plants distributed around the globe and running yearly, the question arises as to what is the use and storage capacity of the captured CO_2? There is enough storage capacity globally in porous, permeable subsurface rocks overlain and underlain by impermeable strata that prevents leakage to hold hundreds of years of CO_2 emissions. Companies also have plans to react CO_2 with

H_2O to make low carbon jet fuel as was demonstrated by Royal Dutch Shell and KLM (see Sect. 4.7.1, first paragraph). The uses of CO_2 to precipitate salable commercial products (e.g., baking soda) and as a pure gas were cited in Sect. 8.7.

Afterword

What we have here is the warning that if global food production falls and populations grow, especially in Africa and Asia, **the planet's food carrying capacity will likely be overwhelmed** whether in 2050 or earlier, a little more than one generation in the future as I write, or most assuredly by later in the century if populations increase as forecast. This will be the scenario unless there are major social and bio- and agro-technological advances that **assure food security for all**, a condition that does not exist in 2020. Efforts to feed malnourished populations include a magnificent effort by the World Food Program that in 2019 helped feed 97 million people (1 in ~9 of the Earth's undernourished or malnourished persons) in 88 countries with delivery of 15 billion rations. For this the WFP was awarded the 2020 Nobel Peace Prize. If deliveries of rations were not hampered or prevented by conflicts, political decisions, or infrastructure obstacles, tens of millions more would have been recipients of nourishing food. The result of not stabilizing and improving global food supplies, food security, and food delivery, plus access to clean and sufficient water could be hunger/water driven migrations by land internally and breaching international borders at all costs [26, 27], and by sea where ships, small and large could be run up close to beaches of food sufficient nations to discharge desperate citizens seeking food to avoid death by famine and starvation at any and all costs.

References

1. IPCC, 2013. Fifth assessment report. In climate change 2013. The physical science basis. Cambridge University Press, Cambridge, pp. 215-315, 29 p. Online. http://www.ipcc.ch/report/ar5/wgl/
2. IPCC, 2018. Special Report. Global Warming of 1.5°C. Summary for Policy Makers. Online. www.ipcc.ch/ar15
3. IPCC, multiple co-authors, 2019. Summary for Policy Makers. In Climate Change and Land. Special report on climate change, desertification, land degradation, sustainable land management, food security and green house gases fluxes in terrestrial systems. Online. www.ipcc.ch/srcc
4. IPCC, 2019. Special Report on the Ocean and Cryosphere in a Changing Climate (srocc). Summary for Policy Makers. Online. http://www.ipcc.ch/srocc/chapter/glossary/
5. Knox, J., Hess, T., Daccache, A. and Wheeler, T., 2012. Climate change impacts on crop productivity in Africa and South Asia. Environmental Research Letters, 7: 1-8. Online. https://doi.org/10.1088/1748-9326/7/3/034032
6. Zhao, C., Liu, B., and 27 co-authors, 2017. Temperature increase reduces global yields of major crops in four independent estimates. PNAS, 114: 9326-9331. Online. https://doi.org/10.1073/pnas.1701762114

References

7. Tigchelaar, M., Battisti, R., Naylor, L. and Ray, D.K., 2018. Future warming increases probability of globally synchronized production shocks. Proc. Natl. Acad. Sci., 115: 6644-6649. Online. https://doi.org/10.1073/pnas.1718031115
8. Scheelbeek, P.F.D., Bird, F.A. and 8 co-authors, 2018. Effect of environmental changes on vegetables and legume yields and nutritional quality. PNAS, 115: 6804-6809. Online. https://doi.org/10.1073/pnas.1800442115
9. UNFAO. Salt Affected Soils. Online. www.fao.org/soils-portal/soil-management/mangement-of…
10. UNFAO, 1985. Annex 1. Crop salt tolerant data. Online. www.fao.org/3/Y4263E/Y4263e0e.htm
11. UNFAO, 2018. The State of the World's Fisheries and Aquaculture 2018. Meeting the Sustainable Development Goals. Rome, 210 p. Online. www.fao.org/state-of-fisheries-aquaculture/en
12. Hoeghguldberg, O., Poloczanska, E., Skirving, W. and Dove, S., 2017. Coral reef ecosystems under climate change and ocean acidification. Front. Mar. Sci., 29 May, 40 p. Online. https://doi.org/10.3389/fmars.2017.00158
13. Alexander, P., Brown, C., Arneth, A., Finnegan, J. and Rounsevell, M.D., 2016. Human appropriation of land for food: the role of diet. Global Environmental Change, 41: 88-98.
14. Wilson, E.O., 2002. The Future Of Life. 229 p., Vintage Books.
15. Willett, W., Rockstrom, J., and 35 co-contributors on the EAT-Lancet Commission, 2019. Food in the Anthropogene: healthy diets from sustainable food systems. 142 p. Online. https://doi.org/10.1016/s0140-6736(18)31788-4
16. en.wikipedia.org/wiki/Vegetarianism_by_country…
17. Watts, J., 2020. Chinese mining firms in Zimbabwe pose threat to endangered species, say experts. The Guardian, Sept. 3. Online. www.the guardian.com/world/2020/sep/03/chinese…
18. Population Reference Bureau, 2020. World Population Data Sheet 2020. Washington, D.C.
19. Bui, M., Adjiman, C.S. and 27 co authors, 2018. Carbon capture and storage (CCS): the way forward. Energy and Environmental Science, 11: 1062-1176. Online. https://doi.org/10.1039/C7EE02342A
20. Global CCS Institute, 2020, Jan. 15. Carbon capture and storage technology on the rise for third year in row, climate report finds. Online. Media releases archive. Online. www.globalinsitute.com/news-media/latest…
21. Chaimin, A., 2019, Direct Air Capture: Recent Developments and Future Plans. Online. www.geoengineeringmonitor.org/2017/07/direct-air-capture…
22. Ishimoto, Y., Sugiama, M., and 4 co-authors, 2017. Putting costs of direct air capture in context. Forum for Climate Engineering Assessment working Paper Series 002, 20 p. Online. ceassessment.org/wp-content/uploads/2017/06/Ishimoto-wps
23. Siegel, R.P., 2018. The Fizzy Math of Carbon Capture. Online. grist.org/article/direct-air-carbon-capture…
24. Fasihi, M, Efimova, O. and Breyer, C., 2019. Technoeconomic assessment of CO_2 direct air capture plants. Jour. of Cleaner Production, 224: 957-980.
25. Levin, K and Lebling, K., 2019. CO_2 emissions climb to an all-time high (again) in 2019: 6 takeaways from the latest climate data. World Resources Institute, Washington, D.C. Online. www.wri.org/2019/12/co2-emissions-climb-all…
26. Podesta, J., 2019. The climate crisis, migration, and refugees. Brookings, Washington, D C., unpaginated. Online. www.brookings.edu/research/the-climate-crisis
27. Rigaud, K.K., de Sherbinin, A. + 9 co-authors, 2018. Groundswell: Preparing for internal climate migration. World Bank, Washington, D C, 221p. Online. www.worldbank.org/en/news/infographic/2018/03/19/

Chapter 9
Sanitation 2050

9.1 Introduction

Sanitation has many facets to consider with the purpose of promoting cleanliness and disease prevention for excellent public health. In 2020, with a world population of ~7.8 billion people, 2.2–2.5 billion do not have access to proper sanitation. As discussed in Chap. 5, sanitation for good public health requires clean water for drinking, food preparation and cooking, and personal hygiene (hand washing stations after toilet use), and an adequate sewage system. The latter includes safe management from toilet to collection (on site conveyance), storage, and transport for adequate treatment and safe end use or disposal of human excreta. The same applies to animal wastes especially at commercial food animal operations. In addition, sanitation requires regular pick up of domestic wastes and animal remains and either using and recycling them or disposing of them so that there is no contact or exposure to humans.

9.2 Open Defecation: A Threat to Public Health—A Solution

There are 50 countries with large and small populations where 10% or more of the populace practice open defecation. According to 2017 data, there are 19 countries with 25% or more of their citizens that went to fields, bushes, forests, water ways, and other open areas to deposit their excreta (Table 9.1). Fourteen of the 19 countries are in Africa. Of the other 31 countries where 10–24% of their populations practiced open defecation, 11 are in Africa. Direct contact with bacteria loaded excreta released into waters or onto land results in diarrheal diseases that cause 1.7

Table 9.1 List of select countries, where during 2017, 25% or more of their populations practiced open defecation, a largely unadvertised public health hazard [1]

Country	Percent	Country	Percent
Benin	54	Mozambique	27
Burkina Faso	47	Namibia	49
Cambodia	32	Niger	68
Chad	67	San Tome Principe	47
Cote d'Ivoire	26	Solomon Islands	54
Eritrea	67	Somalia	28
India	26	South Sudan	68
Kiribati Island	28	Togo	48
Liberia	40	Zimbabwe	25
Madagascar	45		

billion childhood sicknesses and deaths of more than half a million children under 5 years old annually. Diarrhea in adults causes lost employee workdays that harm economies [1, 2]. It is essential that all countries invest to abate the practice with urgency in order to stabilize and improve their public health status and their economies. This is especially true for countries in Africa given the continent's projected rate of population increase (1.34 billion in 2020 to 2.56 billion in 2050) and current lack of safe toilet facilities for many.

The eradication of open defecation will require an investment in educational programs that include latrine engineering and maintenance at the least. This was already cited for Bangladesh where 25% of the country's overall development funds were used to send engineers to villages and towns to explain in non-technical language the health and economic benefits that would be gained by using and maintaining latrines. As noted in Sect. 5.2.1, there was a behavioral change over the course of slightly more than a decade (2003–2015) that resulted in the reduction of a 43% open defecation rate to 1%. A direct economic benefit in Bangladesh was that a family's income rose 16% from US$1154 in 2012–2013 to US$1340 in 2015. Less work days were lost to sickness thus supporting economic development and generating more tax income for the government. The same health and economic benefits can accrue to other countries with an open defecation problem if a Bangladeshi approach serves as a model. Funding for countries to reduce their open defecation rate can come from governments themselves, low cost loans from financial institutions such as the World Bank, regional development banks, grants from international aid agencies, and from NGOs. Obviously, construction of sewer systems, treatment facilities, and treated water distribution networks are necessary especially in increasingly populated urban centers where in less developed and developing countries with large peripheral or edge city populations are not well serviced. A sanitation system with clean water, good sewage systems, and regular domestic waste collection for recycling or secure disposal is the basis for good public health and benefits citizens and ecosystems that in turn contribute to a municipality's/nation's social vitality and economic development, both fundamental to the Earths's human carrying capacity.

In recent years India has had ~620 million people that practiced open defecation. In 2014, Prime Minister Modi instituted an action program of funding for "toilets instead of temples." From 2014 to 2019, 11 million toilets (latrines) were built to service 600 million people (55 people per toilet?) with the hope that open defecation would begin to be eliminated. Although open defection has been lowered, it has not been eliminated ofttimes because the government did not explain clearly to the populace how to use the latrines or how to maintain them and how the sewage could be safely disposed of. In one study of four northern Indian states, the open defecation rate in 2014 was 70% but after latrine construction fell to only to 44% in 2018. Modification of the program educational phase now, in everyday language that emphasizes latrine construction, use, maintenance, and safe excreta disposal is a 'must do' that will help further reduce India's open defecation rate even as its population increases from 1.4 billion in 2020 to 1.7 billion in 2050.

A positive sanitation note is that in India, separation of girls and boys bathrooms and the installation and use of hand washing stations has increased their use and significantly reduced the incidence of diarrheal diseases among school children. This is another area that needs improvement in Africa and other regions that have not yet installed school and municipal hand washing stations.

9.3 Waste Disposal Management

9.3.1 *Landfills*

Given Earth's estimated ~9.9 billion population by 2050, more domestic waste (organic and inorganic) will be generated and more from food production and from the factories that manufacture goods for them. Disposal sites, whether for domestic, agricultural, or industrial wastes have to be located away from and downwind from urban centers so that odors and the noise from trucks transporting wastes to the disposal sites do not pose urban problems. Sanitary landfills are engineered geologically and topographically with attention to making sure that seepage through wastes neither accesses aquifers nor runs off into waterways that people can contact nor onto soils to contaminate them. Waste management protocols dictates that landfills must be underlain by impermeable rock or have an impermeable lining, and for some, a concave shape so that seepage can be captured in a collection bowl at its lowest point that can be emptied periodically for transport to a treatment facility. Protocol further mandates that at the end of daily dumping, waste must be covered with earth materials thick enough to prevent access by animals or insects lest they forage in the waste and carry bacterial/viral diseases to people. Unfortunately, the cover does not prevent the escape of the GHG methane generated from the decomposition of organic waste. At some landfills the methane gas is tapped and used as a fuel but at others, more often than not, it leaks into the atmosphere. Eventually landfills reach their limits and new sites have to be found. This can be a problem especially near cities and mega-cities that will house 6.9 billion people (~70%) of the world's 2050 population. Here, solid wastes from millions to tens of millions of people should be collected, recycled when possible (e.g., glass, metals), and securely disposed of on a regular schedule such as one or two times a week. More of the highly and densely populated cities are evaluating incineration plants for disposal of the masses of wastes they generate.

9.3.2 Incineration

There are more 'pros' than 'cons' for using incineration for waste disposal vs. landfills. A major 'pro' is that waste volume is reduced by 90% or more to ash. Designed properly, incinerators use chemical scrubbers and particulate precipitators to reduce or eliminate pollution from toxic gases (e.g., methane), heavy metals, fine-size particulates (<2.5 μm) and odors. An economic plus is that captured metals can be recycled, as may be ash unless it contains toxicants and needs safe disposal. The EU requires that incinerators reach a temperature of 850 °C (1560 °F) for 2 seconds in order to kill germs and destroy chemicals that might be present. Incinerators are operational year round whereas landfill operations may suffer weather delays in snowy and icy climates. A 'pro' implemented in some European nations is that incinerators are integrated into systems that use steam from the combustion to turn turbines and generate electricity as well as using steam in modern heating systems at close by homes and businesses, and municipal buildings. Sweden burns 50% of waste in such a system and generates 8% of the country's energy needs. This dual use approach will become more important in future decades as populations and wastes they generate increase. The major 'cons' to the use of incinerators is the expense in building the infrastructure to serve the incinerator as well as the incinerator itself and the emission of CO_2 from fossil fuel combustion if it is not captured and used or disposed of (Sects. 8.7 and 8.7.1) unless renewable energy is used to drive the incineration process. The cost not withstanding, Europe contains half of the world's incinerators with the others mostly in North American and Asia. For maximum efficiency, the amount of waste that reaches an incinerator can be reduced by recycling usable wastes (e.g., metals, glass) and by composting organics before transporting them for incineration. Germany and Austria recycle more than 50% of municipal waste. As with landfills, where incineration is used for disposal, the incinerator should be located downwind from existing, growing, or planned population centers.

9.4 Nuclear Waste: A Hazard in Search of Secure Disposal

As we look forward to 2050 and beyond, a problem for human populations worldwide is disposal of nuclear wastes from 437 nuclear power reactors worldwide and from nuclear power facilities that are decommissioned. Nuclear weapons

9.4 Nuclear Waste: A Hazard in Search of Secure Disposal

maintenance and research adds to the nuclear waste disposal problem. There have been accidents at nuclear facilities such as at the Windscale Works, England, (now reprocessing nuclear fuel) and at Three Mile Island in the United States where the public has not been exposed to radiation sickness. But there have been two major accidents we know of where people have been killed and have suffered radiation sickness such as at Chernobyl, USSR (1986), and most recently at Fukushima, Japan (2011, see Sect. 4.4.1). At Chernobyl, post-explosion, radiation within the reactor was initially being contained by an encasement structure of steel and concrete (sarcophagus) but it began failing and has since been replaced by a steel confinement structure built over the sarcophagus. Extensive areas away from the plant are off limits to humans because of life-threatening radiation levels but sustain hearty animal and plant populations. At Fukushima, water fed into the damaged reactor to cool it and prevent a nuclear explosion now passes through carrying radioactive isotopes. The water is captured and is being stored in huge steel containers as isotopes decay. There has been leakage of the contaminated groundwater into the Pacific Ocean. This hurts the Japanese fishing industry as food fish are basically unsaleable to other countries although the amount of radiation is reported as being within acceptable levels.

The problem of nuclear wastes from power plants began 70 years ago with no solution as of 2021 for permanent storage of their radioactive wastes or of wastes after they are decommissioned. At present, they are temporarily stored in cooling pools and in holding tanks and other containers on site or close to onsite. In 2010, there were an estimated 250,000 metric tons of highly radioactive wastes worldwide that need permanent storage sites that can lock them up for the 100,000 years during which radioactive decay should bring them to levels not harmful to humanity [3]. Characteristics of an acceptable storage site include the following: location far away from populated areas or areas designated for population growth; a tectonically stable environment…no earthquake or rupture response as tectonic plates move or volcanic activity during millions of years; disposal in rock that is non-porous, impermeable, resistant to reaction from high heat (e.g., granite) with a thickness that can be developed for permanent storage at depths of at least 250–1000 m for a mined out repository. As countries develop renewable energy sources to reduce the use of fossil fuels that emit CO_2, the use of nuclear energy will continue to increase (see Table 10.2) and there will be more radioactive waste to store but still with no permanent storage locations. Finland, France, Sweden, Canada, and the Republic of Korea are preparing underground disposal facilities that have the characteristics cited above with Finland planning the first deposits in the mid-2020s. The other 29 countries with nuclear power plants are still trying to find their solution to the problem that promises to grow as the Earth's population grows.

As we evaluate the earth's human carrying capacity during this twenty-first century and have to allow that populations near and downwind of nuclear plants are at some risk from a major core melt accident. During the more than 14,816 cumulative operational years of nuclear power plants from 1954 to 2011, the risk of **a major core meltdown** such as happened at Three Mile Island (1979), Chernobyl (1986), and Fukushima (2011) is low, 1 in 3704 reactor years, or ~0.025% [4]. When extrapolated to 2019 with ~17,000 cumulative reactor years of experience, the statistical rate drops to 1 in 5660, or 0.0176%. This does not include a leak or an accident of unknown radioactivity release that is not revealed in full by a country but one that can be detected by analysis of samples of the atmosphere in other countries. During June, 2020 man-made radioactive nuclei ($Cs^{134, 137}$, Ru^{103}, and I^{131}, plus Co^{60}) were detected in the atmospheres of Sweden, Norway and Finland in above normal but still safe concentrations. The I^{131} is the product of a fission reaction. These radionuclides in the atmosphere represent an anomaly (damage?) in nuclear fuel elements and were thought to be related to a leak from one of two Western Russia nuclear facilities, possibly from weapon development. Nonetheless, the reality is that at present, the statistical probability of a core melt down and major radiation release is low.

Although the statistical probability of a nuclear accident is low, it can be improved following basic norms. First, there must be exacting quality control on every piece of material used to build a nuclear power facility and on replacement parts. Second, plants must be built to withstand natural hazards (earthquakes, tsunamis, flooding, loss of normal cooling water), and have double redundant safety controls. Third, there must be high education levels and a yearly psychological evaluation for plant personnel from nuclear engineers to all other employees. Fourth, biannual accident simulations should be practiced that are thoroughly reviewed for flaws in response that lead to fixes of flaws. Fifth, testing of immediate alert alarm of an accident to 'neighbors' and plans for rapid evacuation of populations that could be affected by radioactive fallout. Another option for consideration to reduce the potential reach of an accident is to build small nuclear plants but with the same safety norms as given above (discussed in Sect. 10.1.3, second paragraph).

Afterword

The nations of the world working together have the capability to develop the capacity to deal with waste where it originates and when it has to be safely disposed of, nuclear waste not withstanding. To reach this capacity where sanitation controls are minimal or lacking requires investment, most often in less developed and developing nations. Whether governments in these nations can 'bite the bullet', invest internal funds and receive financial and technical assistance from developed nations and international lending organizations is questionable in some cases. This is because of imbedded histories of 'cultural corruption' and lack of transparency and 'disappearance' of significant funding in past development projects [5]. If these barriers to progress are greatly minimized or ideally eliminated, the potential of receiving significant funding (low cost, long term loans, grants) for projects to improve sanitation programs focussed on waste disposal infrastructure is high. The funding has to include education to build up an engineering cadre responsible for creating and maintaining waste disposal systems and education to raise citizen awareness of the benefits of their use such as better health and improved family incomes. Add to this the promise of better futures for their children sooner rather than later as populations grow to 2050 and beyond. With these aims achieved as a basis, carrying capacity globally with respect to dealing with wastes in the present and future is an attainable goal.

References

1. World Bank, 2019. People practicing open defecation (% of population). Online. https://data.worldbank.org/indicator/SH.STA.ODFC.ZS
2. WHO, 2017. Diarrhea Fact Sheet. Online
3. Geere, D., 2010. Where do you put 250,000 tonnes of nuclear waste? Online. www.wired.co.uk/article/into-eternity-nuclear...
4. Rose, T. and Sweeting, T., 2016. For Safe Is Nuclear Power? A study suggests less than expected. Bull. of the Atomic Scientists, 72: 112-115. Online. thebulletin.org/2016/03/how-safe-is-nuclear...
5. Transparency International, 2020. Corruption Perceptions Index 2019. 29 p. Berlin. Online. https://images.transparencycdn.org/images/2019_CPI

Chapter 10
Natural Resources Beyond Water and Food 2020–2050

What is the status of natural resources availability and access in 2020 other than water and food and how is this projected to change in 2050 and beyond? Among resource categories considered here are energy, critical/strategic metals, industrial minerals and rocks, loss of forest land or no loss as important sources of timber, food, and medicinal plants.

10.1 Energy Resources

Given the estimated growth of global populations in Africa and Asia, and the drive for industrial development, global demand for energy will increase as the century moves from 2020 to 2050 and beyond. To meet the demand while respecting the need to reduce emissions of CO_2 from fossil fuel sources requires more energy generation from non-CO_2 emitting sources and greater efficiency of energy transmission from sources to users.

Estimates of the world's energy outlook in the future can be calculated taking into account several factors such as global warming and prospects to slow and stabilize it, access to energy sources, plus source reliability and affordability as global populations grow and demand increases. Estimates include existing and planned public policy in terms of economic development and energy needs. In a sustainable development scenario, global warming would be arrested before a 2 °C rise but would be best served if the rise were kept to a maximum of 1.5 °C, a possibility only by following the four tined attack described earlier in the text and adhered to worldwide (see Sect. 8.7.1). The latter is questionable given the continued reliance on fossil fuels to power an expansion of industrial development especially in China, India, and South and Southeast Asia.

© The Author(s), under exclusive license to Springer Nature Switzerland AG 2021
F. R. Siegel, *The Earth's Human Carrying Capacity*,
https://doi.org/10.1007/978-3-030-73476-3_10

Table 10.1 The world's projected energy mix 2018–2040. In millions of tons of oil equivalent. Modified [1, 2]

	2018	2030	2040	% of Mix estimated (2030)	(2040)
Oil	4500	4750	4900	29.3	27.7
Natural gas	3500	3900	4500	24.1	25.4
Coal	3850	3900	3750	24.1	21.2
Nuclear	700	800	900	4.9	5.1
Renewables[a]	300	750	1300	4.5	7.3
Modern bioenergy[b]	700	1050	1300	6.5	7.3
Solid biomass[c]	650	600	550	3.7	3.1
Hydro	350	450	500	2.8	2.8
Global total	14,550	16,200	17,700	99.9	99.9

[a]Solar, Wind, Geothermal, Wave
[b]Sustainable Crops For Biofuel
[c]Wood, Crop Residue, Dung, Charcoal

Table 10.1 shows computer model results of a 1% annual rise in energy use globally from 2018 to 2030 to 2040 for energy sources with their percent contribution to the whole [1, 2]. The results predict that during the coming generation, the contributions of oil, coal, and solid biomass will decrease in the projected global energy mix. Conversely, renewables, modern bioenergy, and nuclear will increase with renewables having the highest increase. The hydropower contribution remains the same although thousands of dams are being built or are planned but not all for power generation but also for flood control, to sustain irrigation in times of drought, and as reservoirs for domestic water supply also during times of drought. The addition of new hydropower capability is balanced to some degree as less power will be generated from existing hydropower dams in part because of climate changes in amount, location, and timing of rainfall as well as with less surface water originating from melting mountain glaciers. (Chap. 7, Sects. 7.2.3 and 7.5.2). One would expect a larger decrease in the coal contribution because of less use in industrialized nations in Europe, the United States, and Japan. However this is initially more than compensated for (bump up projected for 2030) mainly from increased coal use in the afore mentioned Asian region. Coal use is expected to decrease in 2040 as more use of natural gas and renewables takes hold in the Asian energy mix.

There is another assessment of the world' projected energy mix more in line with the UN Sustainable Development Goal on Climate Action. Here the plan is to keep the global warming by 2050 under 2 °C by reducing CO_2 emissions to the atmosphere spurred by the use of specific technologies that will reduce them in the effort to meet that target. The figures shown in Table 10.2 differ significantly from those in Table 10.1 (compare 2040 estimates) with the mix estimates more sharply defined in projections for 2050.

10.1 Energy Resources

Table 10.2 The world's projected energy mix 2020–2050 in millions of tonnes of oil equivalent [3]

	2020	2030	2040	2050	% of Mix (Est.) 2040	% of Mix (Est.) 2050
Oil	4554	4528	4070	3664	22.6	18.8
Gas	3355	3719	4115	4522	22.9	23.2
Coal	3769	3709	3462	3159	19.3	16.2
Nuclear	747	874	1135	1657	6.3	8.5
Renewables[a]	673	1210	1894	2457	10.5	12.6
Biomass[b]	1490	1657	1946	2498	10.8	12.8
Hydro	999	1114	1245	1353	6.9	6.9
Others[c]	84	82	108	159	0.6	0.8
Global total	15,671	16,893	17,975	19,469	99.9	99.8

[a]Wind + Solar
[b]Modern Bioenergy + Solid Biomass
[c]Geothermal, Wave

10.1.1 Fossil Fuels

Table 10.2 shows marked lessening in the contributions of oil and coal to the world's projected energy mix from 2030 to 2050 and strong gains during that same period for renewables, biomass, nuclear, and to a lesser degree hydro as energy sources. The decline for oil in the mix would be caused primarily by less demand from the transportation sector because of an increase in the number of EVs, HEVs, and PHEVs (cars, trucks, buses), plus greater efficiency of internal combustion engines. Nonetheless, oil may be down but not out because of its use by industry and as the raw product for petrochemicals.

In 2020, Asia with 60% of the global population was responsible for 75% of coal combustion. The lack of natural gas in Asia and the high cost of importing natural gas as LNG, plus the comparatively low cost of coal (e.g., from Australia and Indonesia) make coal the continent's energy source of choice, especially for steel and cement production in the developing countries the region (e.g., Vietnam, India, China). As technology advances and if carbon capture storage/use systems are applied in Asian industries, there can be dramatic reductions in their emissions and particulate pollutants releases to the atmosphere. In the power generation sector outside of Asia, nuclear facilities and renewables are reducing the use of coal (and oil) so that fossil fuel use will continue to decline.

Natural gas is abundant and extraction is relatively inexpensive. It is easy to transport through supply/distribution pipelines, is a comparatively cleaner fossil fuel than coal and oil, and is available to foreign markets as LNG. Economically and to a degree environmentally, natural gas is a fuel of choice for power plants, industrial projects, and domestic cooking and heating needs. Because of this, natural gas is projected to continue to increase its contribution to the world's immediate and future energy mix.

An additional use for natural gas is being advanced by Aramco is to make a low carbon fuel in a process that uses the natural gas methane (CH_4) as a source of hydrogen (H_2) to react with nitrogen (N_2) extracted from the atmosphere to make ammonia (NH_3). This is the Haber-Bosch reaction that is used to extract 70 million tons of nitrogen from the air annually to make fertilizer. Ammonia can be used as fuel stock added to coal in thermal power plants. In the Aramco process, N_2 is extracted from introduced atmosphere and reacted with CH_4 at high temperature (450 °C) and high pressure (200 atm) catalyzed by iron to give ammonia and CO_2 with the CO_2 captured and sequestered for later use. For the manufacture of 40 tons of ammonia that is being shipped to Japan for a demonstration test, 80 tons of CO_2 was produced. Of this, 50 tons were captured with **30 tons programmed to be used to make methanol** and **20 tons for enhanced (secondary) oil recovery**, thus reducing by >60% the amount of CO_2 that would be produced if natural gas alone was the Japanese fuel stock. For the demonstration test, there will be mixed combustion of 20% of the ammonia by energy content with coal. Positive test results will be the basis for increasing the ammonia input, further decreasing carbon emissions. Japan expects to reduce its CO_2 emissions 26% by 2030 from 2013 levels thus complying with the Paris Agreement with what is designated as 'blue ammonia'. The 'blue ammonia' manufacture uses fossil fuel to generate the reaction energy. There is a plan for Aramco to make a 'green ammonia' that uses only renewable energy (solar and wind) for its manufacture that will eliminate the source energy carbon footprint. Potential users of 'blue ammonia' as fuel stock for power generation would reduce their CO_2 emissions. However, methanol fuel used in internal combustion engines releases CO_2 to the atmosphere. Similarly, use of secondary recovered oil driven by CO_2 gas and refined to gasoline or its use as feedstock in the petrochemical industry would emit CO_2 to the atmosphere unless captured and stored or used. The added CO_2 from these sources, if not captured on site, would not result in a zero sum emission product [4].

Even with the large declines projected for oil and coal in the world's energy mix from 2020 to 2050 as shown in Table 10.2, fossil fuels will still comprise almost 60% of the projected world energy mix. It will be hard to further reduce their contribution to the mix for reasons stated in a previous paragraph. They are abundant, cheaper presently than renewables (e.g., solar and wind), practical, and can be stored and shipped to users when ordered. In addition, fossil fuels emissions of gases and solids will be reduced with improved carbon capture and storage/use/disposal methods during the century. This reduction can be brought about by a political policy decision to impose a costly enough carbon tax that makes it more expensive for an operator to emit than to capture and store. This means either a retrofit or purpose build system for the capture and storage of CO_2. New generation CCS systems will likely come on line during the next decade. It is certain that with the aim of balanced and practical energy options for the future, fossil fuels will maintain their place for use in many industrial applications.

10.1.2 Solar and Wind Electricity Sources

The global shift to **solar and wind energy** to produce electricity as the century progresses is ongoing because costs for materials, installation, and maintenance are decreasing as efficiency in production and transmission of electricity is increasing. For example, from 2010 to 2020, the cost of building solar energy farms dropped 80% and the cost of building large wind turbine farms dropped 40%. The greater population growth in sunnier regions (e.g., Africa, areas of Asia) is creating greater demand for electricity for domestic use, smaller manufacturing ventures, and education (light for students to study by), all of which can contribute to a country's development goals. This demand can be serviced to a significant degree by government and/or utility company investment in small solar installations in rural areas and large solar farms near urban populations spurred by the aforementioned decrease in prices and improved increase in transmission efficiency.

Similarly, the **wind** energy contribution to the electricity grid continues to grow as costs for equipment purchase, installation with necessary infrastructure, and maintenance decline as output increases. In 2009, the cost per MegaWatt-hour was $70 but by 2018 the cost was $20 per MegaWatt-hour. Similarly and worth mentioning is that during the same time frame the solar energy price was reduced from $150 per MegaWatt-hour to the $20 per MegaWatt-hour to match wind energy. This price beats that for natural gas by ~$10 per MegaWatt-hour. Presently, old wind turbines are being replaced by newly designed turbines that increase the power generated into electricity transmission systems.

Future projections have sun-blessed nations and coastal nations creating more solar farms and onshore/offshore wind turbine farms, respectively, to satisfy a growing societal demand for 'green' electricity that doesn't contribute to global warming. As indicated in Tables 10.1 and 10.2, analysis of computer data estimate that these renewables and others (geothermal, modern biomass) will continue to bring about a reduction in power output generated by fossil fuels. In a preview of what this reduction might mean for other nations, solar, wind, and hydropower in the United States generated more electricity than coal on 90 of the first 135 days of 2020 vs. 38 days for the entire 2019. On days with little wind and/or clouds that block out the sun, power plants fueled by natural gas that supplies 38% of the United States electricity inventory can increase electricity delivery to the grid to make up the renewables deficit [5].

10.1.3 Nuclear Power Plants

Nuclear power plants will increase their input to the world's energy mix. There are 55 plants under construction in 15 countries. About half of these are in 5 countries (12 in China, 7 in India, 4 each in the UAE and Korea, and 3 in Russia). Another 100, mainly in Asia, are on order or planned and another 30 countries are considering planning or starting nuclear power programs. This 'green' energy source will contribute electricity to global grids without CO_2 or particulate emissions [6]. The problems of nuclear wastes and the probability of a major accident with a core meltdown such as at Chernobyl and Fukushima were discussed in Sect. 9.4. We may question "should nuclear waste disposal locations be determined before new plants are built?" As noted in Sect. 9.4, third paragraph, five countries are building underground disposal sites but for others, the answer seems to be "we will continue looking for secure disposal sites, but let us keep generating 'green' electricity."

In the future, and as a result of a successful Russian project, electricity and heat can be generated by small modular (nuclear) reactors (SMRs). In a successful pilot project an SMR operating from a barge was floated to a remote region (Chukotka) in the Arctic tied into the Russian grid. The SMR generates electricity and heat to a sparsely populated area and thus facilitated the closure of a coal-fired power plant. The SMRs may be useful as well to supply electricity for sea water desalination plants. The Russians are now planning to build an SMR on land to be operational by 2027. Remote areas and those without access to national grids can benefit by SMRs. Argentine and China are building prototypes of SMRs as models for remote regions of their countries [7]. The SMRs, being small, (generate up to 300 MWe vs >1000 MWe up to ~7000 MWe for most operating reactors) will have a smaller reach of damage in the event of a radiation release.

10.1.4 Biomass

Modern biomass, crops (e.g., corn, sugar cane) grown in many nations to support ethanol production, is sustainable and is projected to increase its role in a future global energy mix, notwithstanding the increase in EVs in the transport sector. Ethanol makes up 10% of gasoline for cleaner operation of internal combustion engines.

10.1.5 Battery EVs, HEVs, PHEVs

In 2019, (battery) electric vehicles (EVs), hybrid electric vehicles (HEVs), and plugin hybrid electric vehicles (PHEVs) captured only 1.4% of the car market in the United States mainly because of a tax credit incentive that was low and a lack of charging station infrastructure on state and national highways. Conversely, Norway provided high subsidy incentives to its citizens, developed a reasonable charging infrastructure, with a result that EVs captured 42% of its car market. Other incentives used in Norway are no fuel taxes (= US$0.50/gal gasoline) and special highway lane (HOV) access. As costs for batteries (hence vehicles) decreases, as battery performance increases to allow improved mileage range on a full battery charge, and as an infrastructure of charging stations are put in place on local, regional, and national routes, the purchase of electric vehicles will increase accordingly.

A realistic assessment of a 2050 contribution to the global energy mix compared with earlier decadal years presented in Table 10.2 shows an overall marked decline for fossil fuels and large gains for renewables, nuclear, modern biomass, and hydropower. In the near future (2035–2050) and later in the century, electric vehicles will take their place in societies as a major transport sector at the expense of internal combustion engine powered vehicles that will greatly diminish their production or ceased to be built by some companies by 2035. The same is not the case for commercial air transport in the foreseeable future.

10.2 Critical/Strategic Metals, Industrial Minerals and Rocks

Critical materials are those essential to industrial processes for the production of goods or components. Strategic materials are those essential to defense and national security needs and those required for economic development and public health and safety. These materials have special properties necessary for industrial production and defense purposes. As important as they are to 2020 societies, they will undoubtedly be more important later in the century as populations grow and demand for products made using them increases. If critical and/or strategic materials have to be imported, they are subject to disruption because of political or economic reasons. In such a scenario, alternate sources have to be available or a domestic supply has to be developed or a substitute has to be found. In 2010, China, with ownership of 90% of the rare earth metals inventory restricted exports to 20% of its production. The 20% was far less than was needed for various uses by the international community [8]. The price of these commodities rose by up to 40%. As a result, a mine in California, United States that had been closed years before because of the low price offered by the Chinese at that time reopened. It is in production. In 2017, the United States had no production rare earth elements but in 2018, 15,000 metric tons were produced. Another high-grade rare earth elements deposit in Colorado is being developed. Other countries seeing the demand and an economic incentive to develop domestic sources for their critical/strategic elements needs or as valuable export commodities have done the same. Each country or country association has its own listing for these materials. Table 10.3 shows a list mainly of metals classified by the United States in 2018 and the European Union in 2017 as critical/strategic.

Likewise, many industrial minerals and rocks fall into the critical/strategic class because they are essential to agriculture and industry. Examples of some are listed with their unique uses in Table 10.4.

The Earth has an abundance of these critical/strategic commodities but not often in the concentrations in rocks that make them viable ores that can be extracted, processed, and sold at a profit for a mining company. Unlike some energy sources,

Table 10.3 List of 35 materials, mainly metals, considered critical/strategic by the United States and 27 by the countries of the European Union. Minerals common to both are shown in bold [9, 10]

Aluminum (bauxite)	**Germanium**	Rhenium
Antimony	**Graphite (natural)**	Rubidium
Arsenic	**Hafnium**	**Scandium**
Barite	**Helium**	Strontium
Beryllium	**Indium**	**Tantalum**
Bismuth	Lithium	Tellurium
Cesium	**Magnesium**	Tin
Chromium	Manganese	Titanium
Cobalt	**Niobium**	**Tungsten**
Chromium	**Platinum Group Metals**	Uranium
Cobalt	Potash	**Vanadium**
Fluorspar	**Rare Earth Elements**	Zirconium
Gallium		

Note: The EU list also includes Borates, Coking Coal, Phosphate Rock, Silicon Metal, Natural Rubber, Phosphorus

Table 10.4 Examples of some critical/strategic industrial rocks and minerals and their uses [11]

Phosphate rocks (fertilizer)
Salt (food preparation and preservation, chemical use)
Sand and gravel (concrete)
Limestone (cement, building stone)
Gypsum (drywall board)
Clay minerals (e.g., kaolin, bricks)
Diamonds (drilling bits for oil/natural gas)
Barite (oil well drilling control)
High purity quartz (silicon): (process equipment for semi-conductor chips, solar cells, high temperature resistance tiles such as used for space vehicle reentry shields)
Kaolin (printed circuit boards)
Microcline (high voltage electrical insulators)
Spodumene (batteries, rocket propellent)
Graphite (nuclear reactors)
Sulfur (chemical manufacture)

ores are not renewable at least in the concept of human time on Earth. This is not absolutely true because there are massive mineral deposits on the deep ocean floor that are being added to as molten matter beneath the floor spews through it and forms masses (chimneys) that contain huge tonnages of many critical/strategic metals. We are not yet at an economic stage that calls for mining them, but the inventory reserve is there and equipment can be put in place together with the environmental protocols that would allow mining and protect associated marine ecosystems.

For a government to plan for the critical/strategic elements and industrial minerals and rocks supply in the future (e.g., 2050 and beyond), it must first review existing demands. It must also assess availability, and cost of a critical/strategic commodity in a politico-economic atmosphere where a major source can be limited, disrupted, or cut off (e.g., rare earth elements, petroleum). Next, governments, especially those with growing populations have to estimate what their future needs for commodities will be. The 2.3 billion more people in 2050 over the 2020 global population will increase product demand, much from growing middle classes mainly in Africa, Asia, and Latin America. Third, governments that import significant percentages of their metal/non-metal needs will have to address how they will meet added demands for the manufacture of products requiring such raw materials. One answer for a government would be to forge trade alliances or barter system with counterparts for what its needs are for what it has that others need. An added answer is that a government will have to finance geological exploration field work to find new sources of commodities that are needed to meet agricultural/industrial requirements. The exploration may be in another nation that is interested in attracting development investment that promises future royalty income and employment opportunities for its citizens. Investors wanting a project 'in' for political as well as exclusive access to a sought for commodity or commodities and also profits can offer added incentives if a project is approved. These may be agricultural assistance, and improvements to infrastructure from sanitation to health clinics to utilities to communication to transportation and other sectors that need modernizing or construction.

Discovery of a potential critical/strategic commodity deposit by exploration geologists is followed by evaluations to determine whether it will be viable economically for exploitation from extraction through preparation/refining stages to use in an environmentally sound project given existing and estimated future pricing. In the end, inventory stocks of these commodities should be available to all users at fair prices but not because of politics directed economic restrictions. As earlier noted, China's restriction on rare earth elements exports had political/economic aims. Nor has it been the case viv-a-vis oil supplies that in recent years and during 2019–2020 has seen production being limited to drive up oil prices, or increasing production to drive prices down and put competition from shale oil/natural gas production out of business. This tact was not long lasting because producer countries can not forego income for extended periods of time during which the competition goes dormant but ready to resume production when the 'price is right'.

10.3 Loss of Forestland or Not? An Important CO_2 Sink, a Source of Timber, Food, and Medicinal Plants

10.3.1 Negative Use Impacts, Positive Use Benefits

Forests cover 4.06 billion hectares (ha) or more than 30% of the Earth's land surface with 54% (2.19 billion ha) in five countries: Brazil, Canada, China, Russia, and the United States. Of the 2.19 billion ha, 45% (0.98 billion ha) are tropical forests followed by boreal, temperate, and sub-tropical forests [12]. Tropical forests are sites of extraordinary biodiversity and are estimated to contain 2/3 of the Earth's land plants and animals [13].

Forests can continue to provide countries with environmental, social, and economic benefits throughout the twenty-first century if forest management directives for sustainability are engaged. However, in some countries trees are being harvested or cleared in an unsustainable way to satisfy demands from growing global populations for **forest and agricultural products**. For example, Brazil contains 60% of the Amazon rain forest: Amazonia. In 1970, this extended over 4.1 million km^2 but by 2018 the rain forest was reduced to 3.4 million km^2, a loss of 709 thousand km^2 or more than 17% for the wood, and land for beef cattle, soybeans, and palm oil plantations. Malaysian and Indonesian controlled areas of Borneo (27% and 73%, respectively), and Indonesia's Sumatra have suffered similar rain forest loss mainly for the wood and land for palm oil plantations. The demand for Amazon and Southeast Asia timber and agricultural land products is being fueled by improving economics especially for more people with disposable incomes added to middle classes in less developed and developing countries as national populations grow. This bodes ill for preserving the sustainability of forests and the benefits they can provide for future generations. Much of the timber that is harvested is illegal, a subject treated in a following paragraph. It must be noted that the COVID-19 pandemic has put a temporary hold on much of the demand for forest-originated products because of slow down in employment and loss of income by great numbers of people.

From 2000 to 2014, the Congo Basin, the second largest tropical rain forest on Earth after the Amazon, lost 165 thousand km^2 (~5%) of forest and woodland. **Here it is hunger that drives 84% of the clearing for small farm holder subsistence cultivation**, a greatly expanding force that sustains life saving food production but at the same time threatens forest existence by the end of the century. Two thirds of the forest loss is in the Democratic Republic of Congo with a high rate of population growth (84 million people in 2020, 137 million by 2035, 195 million by 2050 and 362 million people by 2100). Overall, the 6 countries with areas in the Congo Basin are forecast to have a fivefold population increase by 2100 [14].

Governments allowing or even encouraging the loss of rain forests legally (e.g., in Brazil's 2019 government) or not stopping illegal timbering are losing sources of vegetation that have led to the development of medicines that are now important to citizens health globally. **Fully 25% of all current western medicines are derived from plants found in rain forests even as researchers estimate that only 1% of the flora and fauna in rain forests have been tested for medically active compounds**. The remaining 99% of forest vegetation that can be tested has the great potential to stoke new medicine development to assist treatment for existing diseases or for new medicines that may reduce the threats from epidemics or even pandemics. One plant highlights the potential of the rain forest to fight disease. The *Rosy Periwinkle* is used to prepare the two most effective medicines against childhood leukemia: Vincristine and Vinblastine. These have increased the rate of survival of children with the disease from ~10% to >95%. What potential medically important compounds are we losing as rain forest invasions continue? Only honest forest management can controls timber harvesting and/or forest clearing in order to sustain wood yields and protect forest ecosystems and their surroundings with the great biodiversity they contain. This would serve humans tangible and aesthetic needs, and potentially contribute new medicines to protect human health throughout the twenty-first century.

10.3.2 Forest Loss

Satellite images supported by on the ground observations are used to calculate changes in the global forest cover. As such, the quality of the images, their recency, and experience of those interpreting the images as well as the areas evaluated by ground inspections can affect the numbers generated and reasons given for those recently presented for forest loss or forest gain.

10.3.2.1 Drivers of Forest Loss 2001–2015

A 2018 report based on computer-trained satellites data (high resolution Google imagery) representing a 15 year span of time from 2001 to 2015 attributed forest loss to five drivers [15]. These were industrial agriculture (commodity driven deforestation) that caused 27% ± 5% of the loss at 5 million ha (12.5 million acres) annually. This was followed by forestry with 26% ± 4%, shifting agriculture (small farm holders) with 24% ± 3%, wildfires with 23% ± 4%, and urbanization with 0.6% ± 0.3%. According to the report, illegal harvesting was responsible for as much as 2/3 of the forest conversion.

Also important is deforestation by wildfires, generally wind driven, that are on an increase worldwide as a result of climate change with heat waves and drought or less rain so that vegetation is drier and susceptible to wildfires ignited by lightening or by humans. During 2019, rainforest fires devastated 7 million ha in the Brazilian Amazon. A comparison of the number of wildfires in the Brazilian Amazon during July 2019 of 5318 with those during July 2020 showed a 28% increase to 6803 wildfires. During the latter 2019 to early 2020 wildfires in Australia ravaged more than 20% of the country's forests or 28 million ha and killed perhaps a millions animals.

10.3.3 A Declining Rate of Forest Loss

According to the FAO Global Forest Resources Assessment, deforestation has lessened markedly during recent decades. From 1990 to 2000 deforestation averaged 16 million ha annually and this dropped slightly to 15 million ha annually from 2000 to 2010. There was a significant reduction of deforestation to 12 million ha annually from 2010 to 2015 that dropped even more to 10 million ha yearly from 2015 to 2020. This represents a reduction in deforestation of 37.5% but the total of forested areas continues to shrink. During the respective corresponding time frames, forest expansion went from 8 to 10 million ha and then fell to 7 and 5 million ha [16]. Tropical forest area declined whereas temperate forest area expanded and boreal and sub-tropical forests showed little change during the same time frames.

10.3.3.1 Forestland Lost, Forestland Gained

Geographically, there have been reported declines in forest areas in South and Southeast Asia, South and Central America, and Central, West, and East Africa. Forest expansion was reported from Europe, North America the Caribbean, East Asia, and West Central Asia [15]. According to a 2019 report, the annual rate of tropical rainforest loss between 2014 and 2018 declined to 4.3 million ha. The total forest loss during the same time period was reported as 26 million ha or 260,000 km^2 ($\cong$ to the UK area of ~248.5 thousand km^2) with most of the tropical rain forest lost to industrial agriculture in Brazil, Indonesia, Malaysia, and Africa [17].

10.3.4 Forest Gain?

Whether there really has been a decline in the Earth's total forest area in recent decades has been put in question by a 2018 publication. Researchers used advanced high resolution radiometers on 16 weather satellites to track and compare daily small changes that added up over the course of time to present a "revised picture." The revision reported that global tree growth from 1982 to 2016 more than offset global tree cover losses. The loss in rainforests suggested to many researchers that the Earth's overall tree cover was shrinking. However, the 2018 report indicates that this is not so and that the Earth's tree cover is actually increasing. New tree cover offset loss by 2.24 million km^2 (+7.1% relative to 1982). This new cover is occurring in sites that had been barren of trees…deserts, tundra areas, on mountains, in cities and other non-vegetated land. Global warming raised timberlines in some mountainous regions and allowed forests to creep into tundra areas and also into abandoned farmlands in Russia and the United States. Humans directly caused ~60% of new global tree growth. Net loss in the tropics is offset by net gain in the extra tropics or temperate regions. Bare ground cover decreased by 1.16 million km^2 (−3.1%) mainly in Asian agricultural regions. Forty percent of the forest gain was attributed to drivers such as climate change. Land use change included tropical deforestation mainly for agriculture expansion as a negative factor while temperate zone reforestation (most in China and Africa), afforestation, cropland intensification and controlled urbanization were positive factors. Changes were consistent with mountains gaining tree cover and with many arid and semi-arid areas having lost vegetation cover. Drivers reflect a human-dominated Earth system [18].

With the Great Leap Forward in China during the 1950s, forests were ravaged with subsequent harm to otherwise productive ecosystems and water security. To rectify the grave error of deforestation, China banned logging in 2000 and initiated a grand reforestation program with an investment of $US14 billion. By 2010, the positive results were evident as the forest cover increased by 45 thousand mi^2 (28.8 million acres/11.5 million ha) as a result of 60.7 thousand mi^2 increase in forest cover against a forest loss of 14.4 thousand mi^2. The government hopes to expand the forest cover to 154.4 thousand mi^2 (99 million acres/39.5 million ha) as the reforestation program continues. There is a negative effect of this program. To that end, China is the destination for more than 50% of the global export of timber (legal and illegal) thereby fueling the decline of forest cover in other regions especially in Southeast Asia and Africa [19, 20]. To cut back significantly on China's illegal timber imports, purchasers of wood products from China would have to require certificates that show the legality of the timber used.

10.3.5 Illegal Logging

Illegal logging encompasses logging in protected forests, felling protected species, exceeding licensed quotas, and transporting and buying illegally harvested timber. It is the result of poor governance and lack of enforcement of laws often as a result of corruption. The percent of illegal logging varies greatly among countries. For example, illegal exports of timber from Peru, Bolivia and the Democratic Republic of Congo has been over 70%, whereas illegal timber from Brazil and Indonesia has been 25% and 50%, respectively. However, percentages can be deceptive because the Brazilian and Indonesian total production is so large that their percent of illegal contributions still represent a large mass of illegal wood. It should be noted that in recent years the Brazilian and Indonesian contributions of illegal exports are down from their 2000–2012 highs [21].

In 2018, the global production and trade in forest products was more than US$270 billion, an increase of 10% over 2017 [22]. Illegally logged wood sells at cheaper prices than legal timber and depresses prices by 7–16%. This costs the legal wood industry US$10 to US$15 billion annually. Governments lose US$5 billion in taxes annually. Illegal logging must be combatted in order to help save the forests and their diversity of fauna and flora as well as assuring sustainable future logging with the jobs it offers and economic returns to concessionaires and governments. Legislation in the United States, the European Union, Australia and Japan requiring certificates of legality of origin has been effective in reducing illegal imports such as from Indonesia where the loss of primary forests slowed by nearly 1/3 between 2017 and 2018. However, this has not been the case in developing countries such as China, India, and Viet Nam [21]. In 2016, China's illegal wood imports were worth more than US$2.9 billion, followed by Viet Nam with ~US$625 million. India was the third largest illegal wood importer at a cost of ~US$525 million. A major problem is corruption and lack of transparency at both export and import ends and in between. Forestry management and both an end or great mitigation of illegal logging are the keys to wood availability to growing populations during the course of the twenty-first century.

10.4 Forestry Management

There are three main keys to preserving tropical rainforests, temperate forests, and boreal forests with their ecosystems and diversity of animals and plants to keep them healthy for future generations as the century advances. These are sustainable logging, minimal clearing for encroachment by industrial agriculture, and for urban expansion. These aims can be attained via tough and transparent forestry management backed by legislated enforcement.

10.4.1 Guidelines for Sustainable Logging

Guidelines for sustainable logging found in many publications are listed below with others that are common sense practices [23–25].

- Determine the maximum number of trees that can be sustainably harvested in the forested area
- Within a harvest area, protected species have to be identified and preserved
- Protect pristine areas >1000 ha with connectivity within concession
- Establish where trees can be felled to as to minimize the size of the harvest area while maintaining connectivity among the gaps to allow for natural reseeding or human planting of saplings
- This may be by felling trees in rows leaving terrain open to natural seeding from uncut trees
- Leave potential for stand regeneration and growth or have replanting as needed after inspection for seedlings
- In tropical rain forest especially it should be standard practice albeit time consuming to cut away connecting vines or other entanglements to prevent adjoining trees not in the harvested group from coming down with a target tree
- Do not clear the understory but rather preserve it to maximize conditions for natural reseeding or replanting of saplings as necessary and to limit soil erosion
- In general regrowth of seedlings into financially mature trees will take 60–80 years (e.g., for red oak, white oak, walnut stands) or can grow taller up to ages of 100–120 years

- Former forested areas that have been clear cut should not be inhabited by humans but rather be repurposed as commercially viable tree farms for future controlled logging and support for rehabilitation of plants and animal diversity
- With respect to watersheds, there should be a buffer zone of 100 ft (30 m) from clear cut areas to limit sedimentation and preserve water quality that could otherwise disrupt their ecosystems

10.4.2 Logging Systems

The logging system selected for a concession should best fit the topography, soil types, proximity to waterways and seasonal conditions. This should direct planning to minimize soil disturbance by skid trails and planning for road densities. Skid trails develop where logged tree trunks are dragged by tractors or animals to points of transfer for transport to sawmills. The gouged and loosened soils open pathways for water runoff that erodes soils. Roads for trucks that carry the tree trunks to transfer points or to sawmills can limit erosion from water runoff. This is especially important in steeply sloping harvest zones. Careful planning can minimize such environmental disturbances and preserve soils and the vegetation they sustain for future generations. Better yet to minimize ecosystem harm is the use of moveable cable systems that lift tree trunks from assembly points to trucks on prepared roads to sawmills or to rivers where they are floated to capture and processing locations. The cable systems are costly and take time to set up but greatly reduce environmental damage. They are used in Europe, Western North America and Japan [26]. Lastly, in some areas of Western Canada, helicopters have been used to move tree trunks to preserve the integrity of ecosystems [27]. By minimizing the disturbance of soils, they will support replanting and reforestation with naturally seeded vegetation or with saplings soon after timbering. As time passes, mature vegetation will be there to serve populations as decades pass…80 years for most harvestable trees and far less times for other vegetation that might bear fruits or berries.

10.5 Preserving the Future: Human Resources

The most precious natural resource is the human resource, the people that make our societies function in good times and when under stress. Citizens apply their abilities, their creativity, their social, economic, and manual or technological knowhow, and their capacity to adapt responses to changing conditions. To this end, they can mitigate or suppress the harmful effects of existential challenges and will do so for those that will arise in the future.

These human resources, highly evident during the 2020 COVID-19 pandemic, evolve initially from exposure to family values. These are reinforced by schooling K through high school, through apprenticeship programs in agriculture and the trades, at community colleges, and at colleges and universities. During COVID-19, national populations have been sustained by fellow citizens worldwide maintaining food systems and other necessary supplies as they continue to produce, prepare, pack, load, transport to user, unload, shelve, sell, and where necessary deliver essentials. Water supplies are maintained. Garbage continues to be collected and disposed of. Police and firefighters remain ready to keep order and respond to calls for help. Health services function from EMTs and other ambulance personnel, to orderlies, to nurses, to medical equipment manufacturers and technicians, to nurses and doctors, to those who clean, to those who prepare and bring food in hospitals and clinics. Postal services, communication (computers, television, radio, print media), and transportation services (bus, subways, airlines) are operated and maintained by fellow citizens. Specialists research with great purpose and care to find treatments for those with active afflictions from the coronavirus and to establish, produce, and use tests to find who may be unknowingly carrying it to isolate them and halt transmission from that source. Other medical specialists in infectious diseases by researching 24/7 have in the course of less than a year developed vaccines that can give healthy persons protection against COVID-19. As noted above, not all nations/societies develop the same response to a health crisis because of their stage of development and a lack of a prepared and well supplied cadre of people in the health sector and others such as those cited above that keep societies functioning. This must evolve from within, aiming to strengthen nations' capabilities to cope with their needs in coming decades in light of growing or contracting populations, global warming and climate changes, and economic and political developments.

For developing and less developed countries, the lack of prepared cadre may be the result of a nation's economic capability. The education of a cadre of prepared individuals can be financed by education focused no cost/low cost loans or outright grants from international organization such as the World Bank, and grants from developed country aid programs and NGOs. This means investing funds in free or reasonable cost schooling through high school (international baccalaureate?) that develops proficiency in reading, writing, mathematics, and the use of computers. The teaching of an appreciation of history so that past mistakes are not repeated and of the arts and humanities gives students a solid knowledge base. Beyond this are the educational aims of each individual as he or she envisions how to fit into his/her society. In some societies, university education is basically free and there are few limits as to how many students can graduate in a specific field. This latter fact can be problematic for a country because it can cause an employment glut in a chosen career. In such a case, educated and talented individuals can emigrate to where their skills can be used…a brain drain for a home country. This is to be avoided for the national good in the health sector and others that maintain social stability.

Afterword

Cultivation of the skills and creativity of our human resources can contribute to development and nation's economic stability and growth by judicious use of natural resources. This means sustainability in their rate of use even as global populations grow. Production of renewable natural resources will have to increase to service and sustain future societal needs for the decades to 2050 and beyond. As discussed earlier, this can be achieved with few exceptions through human ingenuity, invention, and intervention. If natural resources decline in availability and access, perhaps because of cost limitations, the rate of their use should be such as to extend their periods of utility while researchers find viable substitutes. In a perfect scenario, natural resources will support development and employment opportunities that will abet a per capita and a country's GDP. An economic reality check for 2020 and future years is considered in the following chapter.

References

1. IEA, 2020. World Energy Outlook 2019. IEA, Paris. Unpaginated. Online. https://www.iea.org/reports/world-energy-outlook-2019
2. CONTEXT: Energy Examined, 2020. The changing global energy mix: 2018 - 2040. Canada Oil and Natural Gas Producers. Unpaginated Online. https://context.capp.ca/infographics/2018/infographic…
3. Cloete, S., 2019. An independent global energy forecast to 2050, to compare with the IEA's WEO 2019. Online. energypost.eu
4. Brown, T., 2020. Saudi Arabia ships low-carbon ammonia to Japan. Ammonia Energy Assn. Online. www.ammoniaenergy.org/articles/saudi-arabia…
5. Plumer, B., 2020. In a First, Renewable Energy is Poised to Eclipse Coal in the U.S. Energy Information Administration
6. World Energy Outlook 2019. Online. www.world-nuclear.org/information/library/…
7. Fisher, M., 2020. Nuclear Power for the Future: New IAEA Publication Highlights Status of SMR Development. Online. www.iaea.org/newscenter/news/nuclear-power-for-…
8. Siegel, F.R., 2015. Countering 21st Century Social-Environmental Threats to Growing Global Populations. Springer Briefs in Environmental Science, 164 p.
9. U. S. Department of the Interior 2018. Final list of critical minerals 2018. Online. www.federalregister.gov/documents/2018/02/16/2018
10. European Commission Publishes New Critical Raw Materials List, 2017. Online. criticalrawmaterials.org/european-commission-publishes...
11. USGS, 2020. Mineral Commodity Summaries 2020. Reston, VA, 200 p. https://doi.org/10.3133/70202434
12. FAO, 2020. Global Forest Resources Assessments. Key Findings. Rome, 16 p. Online. https://doi.org/10.4060/ca8753en
13. Giam, X., 2017. Global diversity loss from tropical deforestation. PNAS, 114: 5775-5777. Online. https://doi.org/10.1073/pnas.1706264114
14. Tyukavina, A., Hansen, M.C., Potapov, V., Parker, D., Okpa, C., Steinman, S.V., Kommareddy, I. and Turubanova, S., 2018. Congo Basin forest loss dominated by increasing small holder clearing. Science Advances, 4:eaat2993. Online. https://doi.org/10.1126/sciadv.aat2993

15. Curtis, P., Slay, C.M., Harris, N.L., Tyukavina, A. and Hansen, M.C., 2018. Classifying drivers of global forest loss. Science, 361: 1108-1111.
16. Keenan, R.J., Reams, G.A., Achard, F., de Freitas, J.V., Grainger, A. and Lindquist, E., 2015. Dynamics of global forest area: Results from the FAO Global Forest Resources Assessment 2015. Forest Ecology and Management, 352: 9-20.
17. New York Declaration on Forests Assessment Partners, 2019. Protecting and Restoring Forests: Progress Assessment. Create Focus (co-ordinator and editor), 94 p. Online. https://forestdeclaration.org
18. Song, X-P., Hansen, M.C., Stehman, S.V., Potapov, P.V., Tyukavina, A., Vermote, E.F. and Townshend, J.R., 2018. Global land change from 1982 to 2016. Nature, 560: 639-643.
19. Lawrence, W., 2011. China's appetite for wood takes a heavy toll on forests. Yale Environment 360 Magazine, p. 1-5. Online. http://3360.yale.edu/feature/Chinas-appetite-...
20. Viña, A., McConnell, W.J., Yang, H., Xu, Z. and Liu, J., 2016. Effects of conservation policy on China's forest recovery. Science Advances, 2: e1500965. DOI: https://doi.org/10.1126/sciadv.1500965
21. Illegal Logging, Global Forest Atlas, 2020, Yale university. Online. https://globaldeforestation.yale.edu/…/illegal-logging
22. FAO, 2019. Forest Products Statistics. Online. www.fao.org/forestry/statistics/80938
23. Mission 2015: Sustainable Logging Practices. Online. https://web.mit.edu/12.000/m2015/2015/logging.html
24. Logging Conservation Practices, 2020. Global Forest Atlas. Online. https://globalforestatlas.yale.edu/forest-use-logging/…
25. Finland Forest Act (1993/1996; amendments up to 567/2014 included) Online. www.finlex.fi/fi/laki/kaannokset/1996/en19961093.pdf. (national level approach)
26. Trzesnlowski, A., Undated. Wood transport in steep terrain. Online. www.fao.org/3/x0633e/x0622e15.htm
27. Agnew, J., 1997. Designing safe cable-logging methods. In Partial-Cutting Safety Handbook. Online. www.for.gov.bc.ca/hfd/Docs/Sil/SIL435-2.pdf

Chapter 11
Economic Realities in 2020 Populations: What Do They Portend for 2050? 2100?

11.1 Introduction

Four principle factors will determine the economic status of the world population(s) during the coming decades of the twenty-first century: (1) global warming and climate changes it causes; (2) population growth in less developed and developing nations; (3) trade balances; and (4) threats to public health. As emphasized in preceding chapters, governments need the political will to invest now to protect their populations in the 2020 decade and those projected for future decades against social (e.g., water/food security), physical (e.g., natural and human-forced hazards), and health threats (e.g., epidemics, pandemics) that can play havoc with their economies if left to providence. Part of an investment would be well spent in education programs for both financially disadvantaged and advantaged populations on how their actions can affect their future economic status and hence that of their governments and by extension national and world economics. Investment in adaptations that can protect societal interests and businesses, hence employment, can benefit all peoples and increase the probability for future global economic stability and progress. What is a certainty is that the **'now'** costs of adapting by enacting and enforcing programs as a **'world government effort'** designed to slow and halt global warming will be greatly less than economic, social, and political costs of playing **'adaptation catch up'** in the future. This has been emphasized in previous chapters.

In February, 2021 the costs of delay in investing in adaptations were quantified for the United States if the government were to wait 10 years to 2030 rather than act in 2020 to eliminate emissions that drive global warming and hence climate change. The cost of a 10 year delay in implementing policies to achieve zero emissions by 2050 will cost the United States $3.5 trillion. This figure was calculated using the annual costs of replacing fossil fuels with clean energy for electricity, transportation, industry, and construction as estimated for the United States by the scientific community. If the government waits until 2030 to implement policies to transform to clean energy, the cost is projected to be 75% greater than if the transformation began during 2020 and will be **$750 billion/year for a 2030 start or ~$8 trillion. This vs. $320 billion/year for a 2020 start or $4.5 trillion** or a **cost loss of $3.5 trillion with a delay**. Other economies will suffer similar delay costs. In the United States, President Biden has set 2035 as the date to reach zero emissions or close to it. In China, President Xi has set 2050–2060 as the date for China to reach zero emissions [1].

The cost figures cited above are further complicated by another February, 2021 paper that analyzed data on how well nations were adhering to their 2016 Paris Agreement nationally determined contribution (NDC) to reduce CO_2 emissions that would cumulatively reduced emissions by 1% annually. Many countries (e.g., in the EU) were not reaching their contribution levels. If this trend were to continue, the computer calculated probability of keeping the global warming to 2 °C was just 5%. To reach a global reduction that would have a reasonable probability of keeping the warming to 2 °C by 2050, the computed estimate was that the reduction of cumulative emissions of the NDCs would have to increase from the 1% to 1.8% annually [2]. This reinforces the importance of the four-pronged attack on global warming especially if the direct air capture plant being built in West Texas functions as designed to be able to extract 900,000 tons or more of CO_2 annually from the atmosphere (see Sect. 8.7.1). If this proves viable, hundreds of like plants can be constructed worldwide that together can be a great boost to slowing and stopping global warming and the climate changes it forces before it reaches 2 °C.

11.2 The COVID-19 Effect

The COVID-19 pandemic in 2020 has wreaked havoc on the world economies with national lockdowns of all but essential businesses and services (e.g., food production, food processing, markets, pharmacies, hospitals/clinics, waste collection, banks, postal services). Governments' debts increased with falling tax revenues as businesses failed and unemployment skyrocketed to rarely seen levels in many countries. Government funds as were available and loans that were taken out (e.g., from the World Bank and IMF (International Monetary Fund) to the poorer nations) were expended to aid the unemployed and to keep businesses afloat as well as to buy medical equipment, to get health care workers to where they were needed and protective gear for them, to reopen and refurbish hospitals that had been closed, and to set up field hospitals, as well as to support research into vaccines and for sourcing vaccines as they became available at the end of December. Richer nations are in better conditions to cope with increase debt, not so for poorer nations. Low and low middle income countries already with debt stress from loans taken out to improve infrastructure and support social programs (e.g., education and health) as in Africa and South America are suffering most by the new borrowing with battered

GDPs and potential default…bankruptcy…unless there is a suspension of interest and principal payments by debtors or better yet, debt forgiveness. How this may affect GDP rankings projected for 2050 discussed later in the chapter (see Table 11.4) is not clear, especially for Brazil (debt at ~2× GDP), India and Indonesia (debt at 75% or equal to GDP) and Mexico (50–75% of GDP). A second wave of virus infections during December midst some relief from lockdowns caused a rise in unemployment globally as severe lock downs return. As the COVID-19 pandemic spread worldwide stock markets suffered large drops in several sectors (e.g., oil, transportation [automobiles, airlines], tourism, restaurants, in store retail) with great loss of demand and hence employment. Since a pandemic low value during March, 2020 for many markets, stocks rebounded to higher values at the at the end of December into 2021 as vaccines for the disease were distributed and vaccinations began with the aim that all economic sectors will function to gradually reach pre-pandemic levels during 2021/2022. The 1918–1920 "Spanish" flu that killed a reported 50–100 million people of a global population of 1.8 billion was a constant threat to humanity for the next 20 years until an influenza vaccine with a desired adaptability to virus mutations was developed with an excellent level of efficacy. This delay was not the case for the COVID-19 virus because of the accumulated knowledge and technological advances a century of learning brought to today's bioscience communities…our **human resources in action**.

In addition to existential threats posed to economies by the COVID-19 virus, long-term threats to the global economies through the twenty-first century will continue to evolve from the often mentioned population growth and global warming with consequent climate changes. Given scenarios for future decades as discussed earlier chapters, several questions that will affect economic futures can be asked. For example, will there be employment in growing populations for those that have not had educational opportunities, and for high school, community college, trade school, apprenticeship program, and university graduates? Will jobs in the future be at earning levels that promise a better quality of life than existed in 2019 for large populations in less developed and developing nations as well as for smaller populations in developed ones? Only targeted and internationally coordinated 'now' planning for coming decades can hope to open the ways to positive answers to these questions.

11.3 Unemployment, Underemployment: A Global Problem

Poverty/extreme poverty globally has been exacerbated by the unavailability or loss of employment because of the use of robotics (automation), increased mechanization, and the outsourcing of routine manufacturing and service jobs to countries with lower labor costs, less stringent environmental laws, and a good export and telecommunication infrastructure. Each country sets its own financial poverty line according to local (urban, rural) purchasing power. The global extreme poverty level has been set by the World Bank as persons living at <US$1.90 international (intl) per day. In addition to the **unemployment** problem, there is also a problem in some countries with **underemployment** when people work at jobs not commensurate with their education and skills. In the case of the unemployed and underemployed, a sense of discouragement can set in to the degree that they may drop out of a country's workforce or may emigrate to where their expertise can be used. In the latter case this is a "brain drain" that is lost to a country's economic development.

In 2019, close to 10% of the world population lived in extreme poverty (Table 11.2). These impoverished citizens live mainly in least developed, low income countries (45.8% of their population) and lower middle income countries (15.5% of their population) [3]. One UN Sustainable Development Goal is to reduce the world's extreme poverty level by 2030 to 3% of the population or ~250 to 300 million people. This would mean that 100s of millions of those formerly out of work would become employed or better employed so as to contribute to productivity and perhaps tax income for local and national economies. Whether this level can be attained and at the least maintained as the 2050 population reaches a projected 9.9 billion people is uncertain. This is because of the unemployment caused by the factors cited at the beginning of this section even as the demand for goods and services for growing populations increases.

11.3.1 Poverty/Extreme Poverty

As described by the UN, the poorest people suffer severe deprivation of basic human needs. They are often hungry because of lack of access to food and safe drinking water. They lack sanitation facilities, suffer from more poor health, do not have adequate shelter, have less access to education, and regularly have no electric lighting at night. The extreme poverty line set by the World Bank at US$1.90 intl or less per day per person is based on purchasing power in nations' economies. Other levels have been set at $3.20, $5.50, and $21.70 per day to compare the percent of the world population living at $1.90 intl with the percent living at higher levels. Table 11.1 shows these data for 2015 [4].

In addition to the World Bank poverty level classification, each country sets an annual income to determine a person's/family's condition with respect to poverty. In the United States, the poverty line in 2020 was $12,760 for a single person, $26,200 for a family of 4, or 10.5% of the population in 2019. In Nigeria, the per capita poverty line for the period between September, 2018 and October, 2019 was $381.74, for more than 40% of the population.

Table 11.1 Percent of the world's population in 2015 living with international dollar incomes per day that may be used to estimate the number living at varying poverty limits with those living with a $1.90 international dollar considered to be in extreme poverty [4]

Int'l dollar value	Percent world population
<$1.90	9.94
>$1.90 but <$3.20	16.29
>$3.20 but <$5.50	19.78
>$5.50 but <$10.00	18.73
>$10.00[a]	35.26

2015—Number living in extreme poverty: 733.48 million world population not in extreme poverty: 6.62 billion

[a]In 2017 the $10.00 was adjusted to $21.70

11.3.2 Extreme Poverty by World Bank Income Levels

The World Bank also examines those living on $1.90 intl per day by classes of annual country income levels. For 2020–2021 these are: (1) low income ≤ US$1036; lower middle income ≥ US$1036 to US$4045; upper middle income = US$4045 to US$12535; and upper income ≥ US$12535 [3]. Table 11.2 gives the percent of world population in those classes living in extreme poverty <$1.90 intl and those living above it at the class upper limit. Thus, for the lower middle income countries, 15.5% of the population is living in extreme poverty and 46.7% in that class living above it but below the next higher class limiting level.

The targets to reach the UN Sustainable Development Goal of reducing global extreme poverty to ~3% of the projected 9.2 billion population in 2040 (~250–300 million people), are the low income and lower middle income classes. Progress was being made to that end as noted earlier where the number in 2018 had dropped to 650 million or ~8.5% of the ~7.6 billion population but the 2020 COVID-19 pandemic is pushing 10s of millions more into (temporary) extreme poverty, a result that may increase into 2021 until vaccines initially being applied in December 2020 reach most of the world population and take hold, and reliable treatment protocols are in play. In practice, the vaccines were applied in most countries first to medical personnel and health providers, firefighters, police and others in the front line of the fight against the disease. Next in line for many countries have been those persons in nursing homes/care centers and those with underlying illnesses. The elderly, 75 years of age in some countries or 65+ in most others are next to receive the vaccination followed by the general public. In theory, with respect to global distribution of the vaccines, they should be applied to global populations with equal availability to all people no matter their economic status. In reality, each country sets or allows its own norms for vaccinations so that there will be differences. This may depend on a nation's infrastructure and healthcare professional capabilities to receive, store the vaccine, and deliver it to the population. The infrastructure will be lacking in some less developed and developing nations that have not invested enough in it because of their economic limitations.

Table 11.2 Percent of the world population in extreme poverty (<US$1.90 intl) at World Bank class income levels for 2019 [5, adapted]

Low income	<$1.90/day 45.8%
Lower middle income	<$1.90/day 15.5% to <$3.20/day 46.7%
Upper middle income	<$1.90/day 2.3% to <$5.50/day 29.2%
High income	<$1.90/day 0.06% to <$21.70/day 21.4%

Given economic trends, population growth, and worldwide threats to life as known in 2020/2021, we may ask whether extreme poverty can soon be essentially a thing of the past? If so, good. But the world is still tasked with approaches to eliminate poverty that may not be categorized as extreme but damning nonetheless…what is the international dollar range that raises a nation's poor out of poverty…US$3.20 intl per day per person, US5.50 intl per day per person? Maybe, but questionable while governments strive for economic development often at the expense of establishing a good quality of life for all their citizenry while at the same time accepting a high quality of life for some preferred groups.

11.3.3 Status of Global Extreme Poverty

In 2018, the 650 million people experiencing extreme poverty were mainly in a UN designated 47 least developed nations (low income: per capita GDP of <$1036 annually) with 32 of the 47 in sub-Saharan Africa [6]. The 650 million represent **half of what it was** a generation in the past, a notable change. The change was the result economic growth such as in Asia (e.g., in India) [3]. According to the World Poverty Clock, in **2030** at least **2/3** of those living on less than US$1.90 intl a day could be in Sub-Saharan Africa if African populations grow as forecast by demographers. The World Bank estimates that the global extreme poverty population is expected to decline to 550 million by **2050**, but of those still living in extreme poverty, as many as **1 in 9** will live in Sub-Saharan Africa. Because the 2020 COVID-19 pandemic forced a greater or lesser slow down in global economies, more of the world's population temporarily dropped back into the extreme poverty class. Taking COVID-19 into consideration, this class was estimated in **October, 2020** to be more than **736 million people** and growing. As economies open up during COVID-19 with personal attention to wearing masks, social distancing, and hand washing, plus improved treatment and a vaccine for all (in 2021?) and people are employed once again, the number of citizens in extreme poverty will decline as it had steadily done in recent decades. In **August, 2020**, an estimated 318 million people (45%) of the **705 million people** worldwide in extreme poverty lived in ten countries in Sub-Saharan Africa (Table 11.3) [7].

It should be noted that the least developed countries face serious structural obstacles to a sustainable development end such as corruption, autocratic/despotic/

11.4 Global Economic Power Shift

Table 11.3 Change in the number of citizens living in extreme poverty in ten Sub-Saharan nations from 2018 [7] to the start of 2020 [8]

	2018 (million)	2020 (million)
Nigeria	86.9	102.1
D.R. Congo	60.9	66.0
Ethiopia	23.9	36.4
Tanzania	19.9	28.7
Mozambique	17.8	19.1
Kenya	14.7	8.0
Uganda	14.2	19.6
South Africa	13.8	16.4
South Sudan	11.4	11.6
Zambia	9.5	10.4

dictatorship regimes vs. a more liberal governance, inefficient allocation of resources, limited freedom of the press and information exchange, non-independent judiciary, and lack of opportunities to educate and train human resources. As such these nations are susceptible to sharp disruption from economic downturns and environmental disruptions that push people into extreme poverty. The 2020 population of the previously cited 47 least developed countries was ~1.1 billion people or about 13.7% of the global population. The projected population for these countries in 2050 is ~2 billion people and this is estimated to represent ~19.8% of the projected 2050 global population or every fifth person on the planet [9]. Unless the structural barriers to development are greatly reduced in these countries, greater numbers of people will be living in extreme poverty and not contributing to national development and improved economies.

11.4 Global Economic Power Shift

Economic projections made in 2016 estimated that by 2050 the seven largest emerging markets and developing nations (China, India, Indonesia, Brazil, Mexico, Russia, Turkey) could grow on the average at a rate of 3.5% annually through 2050 versus 1.6% for advanced G-7 nations (Canada, France, Germany, Italy, Japan, the UK, and the US). This reflects a gradual economic power shift towards Asia. The shift is towards large and growing populations in the former group (excluding Russia) and stable or contracting populations in the latter group. Table 11.4 lists the nominal GDP values for the top ten nations in 2019 and values projected for 2050.

Although China is predicted to have the highest country GDP by 2050, China's per capita GDP (US$37,571) is lower than that projected for the United States, Germany, the United Kingdom, Japan, and Russia (US$89,878, US$76,250, US$72,972, US$64,151, and Russia $37,775, respectively) [11]. These latter projections for 2050 are an effect of their lower populations with respect to that of

Table 11.4 GDP economic power rankings in trillions of US dollars at MER[a] of the top ten nations for 2019 and that projected for 2050. **Parenthetical values** for 2050 take into account a projected Purchasing Power Parity[b] between nations [10, 11]

2019 Est. [7]	GDP	2050	GDP [9]
United States	21.4	China	49.8 (58.5)
European Union	18.7	United States	34.1 (34.1)
China	14.1	India	28.0 (44.1)
Japan	5.2	Indonesia	7.3 (10.5)
Germany	3.9	Japan	6.8 (67.8)
India	2.9	Brazil	6.5 (7.5)
United Kingdom	2.7	Germany	6.1 (6.4)
France	2.7	Mexico	5.6 (6.9)
Italy	2.0	United Kingdom	5.4 (5.4)
Brazil	1.8	Russia	5.1 (7.1)

By 2050, European Union, France, and Italy have dropped out Indonesia, Mexico, and Russia have moved in

[a]MER, market exchange rates, balance the demand and supply for international currencies
[b]Purchasing Power Parity is a macroeconomic metric that compares economic productivity and standards of living between countries through a 'basket of goods and services' approach when this takes into account a cost if a country used an exchange rate equivalent in $US

China and the probability of steady development vs. the expected good development rate of the Chinese economy. It is impressive that the predicted 2050 Chinese per capita would represent an increased of about 400% over the 2019 level of US$9600. It is important to reiterate that the per capita GDP of countries are averages and not representative of incomes of all citizens (Table 11.3).

While the power shift to emerging markets and developing nations from the G-7 nations as measured by GDP is predicted by mid-century, the predictions for the African continent's socio-political future are dire indeed when evaluated in light of **growing populations, effects of climate change, water/food security, sea level rise, and natural disasters**. For example, the World Meteorological Organization predicts that by mid-century cereal crops will be reduced by 13% in West and Central Africa, 11% in North Africa, and 8% in East and Southern Africa seasonally reducing food security for the 1.34 billion Africans in 2020 and for projections of 1.89 billion in 2035, and 2.56 billion in 2050. This does not include the global reach of other factors that may upset their economies. A recent report estimated how global warming fueled disasters for **temperature rise scenarios** of 1 °C through 4 °C could **decrease** Africa's annual percent change in **GDP**. For example, this would be −5.01 ± 3.30% for a 2 °C rise (realistic) to −12.12 ± 7.04% for a 4 °C rise (an undesirable possibility for the continent and the planet). The Western African region including areas of the Sahel shows the highest reductions in annual percent GDP by approximately double the other regions at −9.79 ± 1.35% for a 2 °C rise and −22.09 ± 2.78% for a 4 °C rise [12]. African regions and financially poor nations elsewhere will need help to apply adaptations to minimize the effects of global warming on their economies.

11.4 Global Economic Power Shift

An important question is whether developing nations and emerging market nations with increased GDPs forecast to average 3.5% annually to 2050, will allocate a portion of the increased incomes for adaptation projects that will improve what is of maximum importance for their citizens as well as a portion for economically stressed regions such as those in Africa. As discussed in earlier chapters, such adaptations are those that will reduce CO_2 and other GHG emissions in the atmosphere in order to slow and stabilize global temperatures and thus attenuate increases of damaging effects by climate changes. Investments made by the nations benefitting from the annual GDP increases for adaptations 'at home and overseas' early on during a three to four decades window from 2020 to 2060 to protect citizenry and support economic activities and hence political stability will be much, much less than waiting and literally being forced to invest later in the window (see Sect. 11.1, second paragraph).

The question cited above remain the same but the predicted country/regional GDPs annual increases have been put in question as a result of the COVID-19 virus that has infected more than 111 million people worldwide, killing more than 2.5 million as of February 19, 2021. A preliminary report presents estimates of the effects of the pandemic on world GDP and exports in two groups: those that are expected to result from a **contained global pandemic** and those that are expected from an **amplified pandemic** if there are delays in containing the disease [13]. **The contained pandemic designation is applied if countries bear only one-half of the economic impact of the full China shock from COVID-19. The amplified pandemic designation is applied if the shocks are uniform across all countries.** The latter is the basis used for figures presented in quarterly reports by the IMF World Economy Outlook. The January, 2021 report, amid significant uncertainty, shows world growth (output) at −3.5% for 2020, +5.5% for 2021, and +4.2% for 2022. Growth in advanced economies shows −4.9% for 2020, +4.3% for 2021, and +3.1% for 2022. Growth in emerging markets and developing economies show −2.4% for 2020, +6.3% for 2021, and +5.0% for 2022. It must be noted that figures such as these change somewhat in each quarterly report or among organizations but the projections for losses and gains are reasonably similar. **For example, the IMF World Economy Outlook for April, 2021 shows adjusted world growth at -3.3% for 2020, +6.0% for 2021, and +4.4% for 2022 with growth in advanced economies given as -4.7& for 2020, +6.1% for 2021, and +3.6% for 2022. Emerging markets and developing countries growth are given as -2.2% for 2020, +6.7% for 2021, and +5.0% for 2022.** Similarly, a January 5, 2021 press release from the World Bank for growth for **all advanced economies shows −5.4% for 2020 and +3.3% for 2021whereas emerging markets and developing countries show a −2.6% in 2020 and +5% for 2021**. The brunt of the contraction in GDP and trade affect those services sector workers most at −9.3% job/income loss, with agriculture and manufacturing suffering a −3% loss. Estimates for 2021 are for growth of +4.8% and +5.9%, respectively, for the two groups [14, 15]. **The GDP and trade numbers will change as vaccinations that began at the end of December, 2020 are applied to the world population and as treatments for the disease are approved. This will result in the world's economy recovery starting in 2021.**

A 'red flag' is waving about preparedness in the future to deal with threats to national, regional, and worldwide health, threats that have a negative impacts on economies. The world was not prepared to respond to the COVID-19 virus for three principal reasons. First, there was the delay by the Chinese government in reporting to WHO and hence the world governments of the disease that may have originated in Wuhan, China and of its potential for rapid spread from human to human. Sadly, this was a repeat of the Chinese response to SARS in 2002. The delay gave the virus a head start in spreading around the world as travelers from the disease area were unwitting vectors of the virus. Second, the city, markets, and laboratories in Wuhan that may have a clue to the virus origin were not initially open to WHO medical experts to visit and help with determining the origin of the virus…animals, humans…when the reality of pandemic was recognized. Third, the health systems in many nations were not initially prepared with testing kits to identify the infected and the healthy, and protective gear for health workers as they treated the scores of millions stricken with the disease. Lastly, the Chinese government held back on the distribution of research results from their own scientists results into the origin of the virus. Good scientific communication and the dissemination of reliable results can help rein in the effects of a pandemic in the future whether from China or other nations or regions in order to ease economic disruptions worldwide.

The probability is real that another pandemic will develop in the future as growing human populations encroach on animal habitats and there is closer contact as virus carrying animals forage in a shared ecosystem. Even as the world's governments forge agreements designed to stem global warming and climate changes that threaten people and property, world leaders must plan now to be able to prevent the spread of a (future) disease from an epidemic to a pandemic stage. This can be accomplished by being prepared to respond to initial disease events from immediate true first reports (track and isolate), by allowing expert collaboration at the beginning of an event (find origin and isolate it), and by being prepared with protective gear for front line health workers as they tend the initial wave of afflicted persons. As we have seen, there have been medical specialists in laboratories world wide that have been developing treatment protocols and in less than a year have created vaccines to protect the global populations against the COVID-19 virus, a feat that is to be honored. This is especially true when we remember that the influenza virus in 1918–1920 killed 50–100 million people and that it took years until a mixed flu virus vaccine was very successful in protecting the public, a mixture experts modify each year to further protect us from the ravages of the disease suffered a century ago. Preparedness as cited above and a well educated cadre of treatment and vaccine development specialists support global economic interests by protecting public health and keeping society functioning at a reasonable level. This should result in future steady or increasing global employment and productivity, and satisfactory per capita and national GDPs.

References

1. Energy Innovation, 2021. The costs of delay. Online: energyinnovation.org/…/2021/01/Cost-of-Delay.pdf
2. Liu, P.R. and Raftery, A.E., 2021. Country-based rate of emissions reductions should increase by 80% beyond nationally determined contributions to meet the 2°C target. Communications Earth & Environment, 2: Article number 29
3. Serajuddin, U. and Hamadeh, N., 2020. New World Bank country classification by income level 2020-2021. Online. https://blogs.worldbank.org/opendata/new-world-bank-…
4. Roser, M. and Ortiz-Ospina, E., 2013 with 2017 and 2019 updates. Global Extreme Poverty. Our World Data. Online, 80 p. https://ourworldindata.org/extreme-poverty
5. Weller, C., 2017. The World Bank released new poverty lines. Online. www.businessinsider.com/world-bank-released-new-pverty-lines
6. United Nations, Committee for Development Policy, 2019. The Least Developed Country Report 2019. New York, 163 p.
7. Ojekunle, A., 2018. By 2030, most of the poorest people in the world will be living in Sub-Saharan Africa. Business Insider. Online. www.pulse.ng/bi/politics/world-poverty-clock-by…
8. world poverty clock. Online. https://worldpoverty.io/index…
9. Population Reference Bureau, 2020. World Population Data Sheet 2020. Washington, D.C.
10. IMF, 2019. World Economic Outlook (WEO) Database 2019. Online. www.imf.org/…/world-economic-outlook-october-2019
11. Hawksworth, J., Andino, J. and Clarry, R., 2017. PwC, 2017. The Long View: How Will The World Economic Order Change By 2050. (The World in 2050). PricewaterhouseCooper, UK, 72 p.
12. World Meteorological Organization, 2020. Multi-agency report highlights the current and future state of the climate in Africa. Online. https://public.wmo.int/en/media/news/focus-africa-launches…
13. Maliszewska, M., Mattoo, A. and van der Mensbrugghe, D., 2020. The Potential Impact of COVID-19 on GDP and Trade: A Preliminary Assessment. (April). Policy Research Working Paper 9211, World Bank, Washington, D.C., 24 p. Online. https://documents.worldbank.org/curated/en/…
14. IMF, 2021. World Economic Outlook (WEO) Update, January, 2021. Policy support and vaccines expected to lift activity. Online. www.imf.org/…/2021-world-economic-forecast-update
15. IMF, 2021. World Economic Outlook (WEO) Update, April, 2021. Managing diverse recoveries. www.imf.org/…/2021-world-economic-forecast-update

Epilogue

The Earth's human carrying capacity may be pushed beyond its limits on multiple fronts to sustain a 2050 population much less an end of century population. As reiterated in the text, two most important of these are population growth, especially in Africa and Asia, and the effects of global warming and climate change on water availability and food production. Certainly, natural resources depletion and need to monitor public health preparedness including the necessity to properly dispose of disease causing wastes from multiple sources are essentials in the efforts to support population sustainability. This book has highlighted problems of the Earth's carrying capacity and offered pathways to their solutions. Solutions take time to take hold so that delays in planning and implementation must be minimized, adaptation actions put into place, and activated with urgency because what nations do now will determine if they will be able to sustain their populations at a good quality of life level in the future. This is especially important for less developed and developing nations but will directly affect developed nation as well. Delay in implementing adaptations will be responsible for sickness and death for great numbers of people and cause socio-economic instability for many nations and the world as we know it. To now, 2021, national/regional economic and political interests of many leaders have put aside the advice of experts on responding to threats to the Earth's human carrying capacity that when greatly mitigated or eventually eliminated would benefit their populations and all world citizens. Many other leaders are striving to follow expert advice but without the urgency that would likely have a negative impact on their economies. Without doubt, however, it is now that all leaders worldwide should demonstrate an adherence to morality, the sense of a common purpose for the good of all countries and all Earth's citizenry. This means choosing to instigate, without untimely delay, plans to advance the Earth's human carrying capacities that can sustain and better quality of life for generations to come.

Index

A
Acidification, 95
Adequate standard of living, 68
Agricultural chemicals overuse, 29
Agricultural crop, 23
Agricultural/industrial projects, 83
Agro-biodiversity, 75
Agro-climatic zones, 39
Air-captured CO_2, 46
Alps, 73–75
Andes, 75
Apprenticeship programs, 60
Aquifers, 80, 81
Arable land, 22
Arable soils, 23

B
Biodiversity, 30–32, 34
Biofuels, 61

C
Calcium carbonate ($CaCO_3$), 95
Calm winds, 77
Carbon capture storage (CSS), 103
Carbon dioxide (CO_2), 71, 72
  'contaminant' problem, 56
  emissions, 26, 34, 43, 46, 63
  emissions/reduction plans
    CO_2 removal from atmosphere, 46–47
    Earth's rate of temperature rise, 45
    economic development, 45
    greenhouse gas, 46
    vehicle CO_2, 46
Chemical scrubbers, 45
Chemical sprays, 44
Climate changes
  in growing zones, 44
  negative effects, 60
Climate zones, 78
CMIP5 models (Coupled Model Inter-comparison Project Phase 5), 22
Coastal cities, 80, 81
Conservation agriculture, 25
Continental glaciers, 72
Continued global warming, 72
Coral, 95
Cosmic (ionizing) radiation, 30
COVID-19 pandemic, 45, 84, 87, 130, 136, 137, 143, 144
Critical natural resources, 66
Critical/strategic metals, 65
Crop management practices, 28

D
Deforestation, 31, 32
Desalination
  definition, 10
  disposal of brine, 11, 12
  ecological problems, 13
  extremely high water stress, 11
  global output, 11
  high water stress, 11
  pipeline, 11
Direct air capture (DAC), 104
Drinking water, 71
Dry regions, 77

© The Author(s), under exclusive license to Springer Nature Switzerland AG 2021
F. R. Siegel, *The Earth's Human Carrying Capacity*,
https://doi.org/10.1007/978-3-030-73476-3

E

Earth's carrying capacity
  biodiversity, 3
  definition, 1
  economically advantaged/disadvantaged group, 4
  ecosystems, 1
  global fertility rates, 1, 2
  global population, 3
  global population socio-economic structure, 2
  human, 1, 3
  hunger, 3
  limitations, 1
  malnutrition, 3
  sanitation systems, 4
  World and regional populations distribution, 2
Earth's ecosystems, 59
Earth's human carrying capacity, 53–55, 60
Earth's magnetism, 64
Earth's societal carrying capacity, 59
Earth's temperature, 45
Ecosystems, 72
Education program, 60
Educational programs, 51
Egypt's Water Lifeline, 82
Electrical conductivity (EC), 93
Electricity, 61, 62, 64
Electric vehicles (EVs), 122
Energy
  biofuels, 61
  countries and electrical production in 2019, 61
  demand, 63
  with electricity, 61, 62
  fossil fuel natural gas, 61
  fuel sources, 62
  hydroelectric dams, 61
  hydropower, 61, 62
  national/regional access, 62
  oil-rich Venezuela, 62
  sources, 61
  use, 63
  world total primary energy supplies, 61
Energy resources
  battery, EVs/HEVs/PHEVs, 122
  biomass, 122
  CO_2, 117
  fossil fuels, 119, 120
  global warming, 117
  nuclear power plants, 118, 119, 121, 122
  solar/wind electricity sources, 121
  world's projected energy, 118, 119

Environmental awareness, 60
Ethiopian Dam Threat, 82
European Geosciences Union, 75
European rivers, 74
Extreme climatic conditions, 44
Extreme natural disasters, 40
Extreme weather disasters (EWDs)
  African nations, 42
  on food supplies, 41, 42
  impact on food production, 40, 41
  insect breeding conditions, 42
  negative effects, 43
  on food supplies, 41, 42

F

FAO estimates, 80
Fixed latitudes, 78
Food animal wastes
  African swine flu infected pigs, 52
  cleansing, 52
  fecal waste, 51
  fermentation approach, 53
  global problem, 52
  improved fermentation method, 52
  internal temperatures, 52
  problem of animal waste, 51
  sanitation, 51
  slaughter house wastes, 53–54
Food insecurity
  CMIP5 models, 22
  enough food, 22
  moderate, 21
  poverty condition, 22
  regions, 22
  severe, 21
Food loss, 27
  agricultural, 27
  reduction, 32
  transport in warm seasons, 27
Food production
  genetic modification, 28
  increasing food production, 29
  nutritional value, 28
  and security, 43
  traditional hybridization, 28
Food production and security, 22
  biodiversity, 87
  challenges, 88
  climate changes
    climate events, 89
    decreased food production, 89–92
    fisheries, 93, 94

Index

food fish, warming temperature
  effects, 94
global food supply, 89
potential source, 92, 93
survival immigration, 88
coal, exports, 100, 101
CO_2 extraction from air, 104, 105
cultivation, 87
feed growing earth populations
  diet change, 96
  healthy diet, 97, 98
  kitchen prepared insects, 99
  plant-based diets, 98, 99
food fish supply/security, warming
  temperature effects, 95, 96
global temperature, 99
industrial plants/exports, 100
regional temperature effects, 101–103
reverse global temperature, 103, 104
2020–2050 warming, 88
vulnerable populations, 87
Food security, 75, 88
  cost, 25
  definition, 21
  dependency on food systems, 27, 28
  erosion and nutrient replenishment, 26
  and food production
    (*see* Food production)
  food systems, 27, 32
  population challenge, 26–27
Food system/production/consumption
  edible food, 33
  expansion of farmland, 30–31
  field studies, 33
  food security dependency, 27, 28
  GE crops, 35
  laboratory experiments, 33
  population growth and farmland
    expansion, 31–32
  in rich countries, 32
  rural land area, 34
  space exposure to cosmic radiation, 30
  speed breeding, 29, 30
  unnecessary waste, 33
  use for ethanol, 34
Food waste, 27, 28, 32
Forestry planning, 68
Forests, 66, 67
  for CO_2 emissions, 67
  ecosystems, 66, 67
  preservation, 68
  raw material, 67
Forewarning, 80
Fossil fuel natural gas, 61

G

Genetically engineered (GE) crops, 35
Genetically modified (GMO) seeds, 44
Glacial ice thinning, 75, 76
Global economic power shift, 141–144
Global food security, 71
Global populations, 71
Global warming, 28, 76, 83, 135
  and climate change, 43
  CO_2 concentrations, 40
  Earth's mean temperature, 39
  negative effects, 60
  Paris Agreement target, 40
  rate of emissions, 40
  regional water availability, 77–79
  warming temperatures, 40
GMO cropping, 35
Gordian knots, 45
Greenhouse gas (GHG) emissions, 22, 46, 71

H

"Heat pandemic", 46
Herbicides, 29, 33, 35, 44
High latitudes regions, 78
Horse latitudes, 77, 81
Human influence, 31
Human natural resources, 61
Human originated waste, 54
Human resources
  Earth's human carrying capacity, 60
  Earth's societal carrying capacity, 59
  education programs, 60
  working together, 60
Hurricane driven storm surges, 80
Hybrid electric vehicles (HEVs), 122
Hydroelectric dams, 61, 71
Hydro-political tension, 72
Hydropower, 61, 62, 75, 81

I

Ice sheets, 72
Incineration, 112
Industrial minerals, 66, 67
Industrial rocks, 66, 67
Intermediate wheatgrass, 25
Invasive vegetation (weeds), 39, 44
IPCC, 81

K

Kernza® kernels, 25
Köppen climate zones, 77, 78

L

Land degradation, 43
Latin America
  deforestation, 32
  ecosystems, 32
  population growth, 31
Latitude zone map, 79
Latitudinal divisions, 77
Latitudinal regions, 77
Locust infestations, 42

M

Macro-nutrients, 23
Malnourishment, 24
Malnutrition, 22
Mass of CO_2, 47
Mediterranean region, 81
Melting-ice thinning process, 75, 76
Melting glaciers, 72
Meltwater flow
  Alps, 73–75
  Andes, 75
  mountain glaciers, 72, 73
Metals
  concentration, 65
  to industrial/manufacturing projects, 64
  iron (Fe), 64
  mining industry, 64
  multiyear process, 64
  oil producing nations, 65
  rare earth metals, 65
  raw materials, 64
  surface earth materials, 64
  traditional oil producers, 65
Micro-nutrients, 23
Minerals, 66, 67
Moderate food insecurity, 21
Modified seeds, 29, 30
Mountain glaciers, 72, 73, 76–77
Multiple swarms, 42
Mutations, 29, 30

N

Nationally determined contribution (NDC), 136
Natural gas, 61–63
Naturally released toxins, 54
Natural resources
  critical/strategic metals, 123
  forestland, loss of
    declining rate, 127, 128
    forest gain, 128
    forest loss, 127
    illegal logging, 129, 130
    negative use impact/positive use, 125, 126
  forestry management, 130, 131
  human resources, 131–133
  industrial minerals/rocks, 123, 125
Nitrogen fertilizer, 29
Non-CO_2 emitting energy sources, 61
Nuclear power plants, 55, 74, 75
Nuclear reactors, 74
Nutrient deficiencies, 29

O

Open defecation, 50, 51
Organic matter, 33

P

Peak discharge, 73
Perennial grain, 25
Pesticides, 29, 33, 44
Pipeline transfer, 80, 82, 83
Plugin hybrid electric vehicles (PHEVs), 122
Pollutants
  CO_2, 56–57
  definition, 54
  Earth's human carrying capacity, 54
  outdoor and indoor air pollutants, 54
  potentially toxic metals, 54
Population growth, 68, 69, 89, 97, 98, 135
Populations, 135
Potable water, import/transference
  aqueducts, 16
  fresh water, 17–19
  pipelines, 14
  transport, 15
  water shortages, areas suffering, 17
Poverty, 68
Practiced open defecation, 49

R

Radiation, 55, 56
Radioactive wastes, 55
Rainfall deficit, 81
Rainwater, 76
Rare earth metals, 65
Reasonable proximity, 80
Red Sea, 80
Reforestation, 66
Regional water availability, 77–79
Renewables, 61, 62, 64

Index

Representative Concentration Pathways (RCPs), 90
Reservoir dams, 71

S
Safety net, 24
Sanitation, 49, 50, 53, 56, 57
   clean water, 109, 110
   nuclear waste
      basic norms, 114
      Chernobyl, 113
      major core meltdown, 113
      man-made radioactive nuclei, 114
      nuclear facilities, 113
      power plants, 113
      sarcophagus, 113
   open defecation
      behavioral change, 110
      countries, 110
      diarrhea, 110
      eradication, 110
      financial institutions, 110
      health and economic benefits, 110
      latrines, 111
      positive sanitation, 111
      rate, 111
      sanitation system, 110
   waste disposal management
      incineration, 112
      landfills, 111, 112
   world population, 109
Satellite analyses, 78
Scandinavia, 78
Sea level rise, 76, 80
Seasonal glacial meltwater, 75
Severe food insecurity, 21
Single water pie, 81
Slaughter house wastes, 53
Small modular (nuclear) reactors (SMRs), 122
Socio-economic development, 71
Soils
   active management programs, 24
   arable, 23
   macro- and micro-nutrients, 23
   maintaining nutrient contents, 25–26
   minimizing erosion, 24–25
   nutrient content, 23
   perennial grains, 25
Speed breeding program, 29
Stable ecosystems, 59
Starvation, 22

Storage locations, 72
Sub-Saharan Africa, 81
Sudanese water supply, 82
Sustainable management, 25

T
Temperature conditions, 71–72
Temperature rise, 75
Tidal flooding, 80
Timbering, 66, 67
Trade balances, 135
Traditional oil producers, 65

U
Undernutrition, 22
Unemployment/underemployment
   extreme poverty, World Bank income levels, 139, 140
   global extreme poverty, 140, 141
   poverty/extreme poverty, 137, 138

V
Vehicle CO_2 emissions, 46

W
Warmer temperature, 44–45
Warming agricultural zone, 44
Warming ecosystems, 39
Warming temperatures, 40
Wastes
   fecal waste, 52
   food animal (*see* Food animal wastes)
   radioactive, 55
   and water chemistry, 50
Water
   cycle, 7, 8
   deficiency, 79
   desalination, 10
   groundwater, 8
   management, 79
   needs, 11
   potable, 14
   reuse wastewater, 13, 14
   at risk/availability/food/hazards
      Ethiopian Dam Threat, 82
      FAO estimates, 80
      rainfall deficit impact, 81
      threats to coastal cities and aquifers, 80, 81

Water (*cont.*)
  twenty-first century progresses, 79
  water deficiency, 79
  water management, 79
  water stress, 79
 stress, 8, 9, 72, 79
 supply/security, mountain glaciers, 76–77
Water to water-starved populations, 82–84
Water towers
 assessment, 72
 glacial ice thinning, 75, 76
 melting glaciers, 72
 meltwater flow
  Alps, 73–75
  Andes, 75
  mountain glaciers, 72, 73
 mountain glaciers, 72
 peak discharge, 73
 storage locations, 72
 vulnerability, 72
Weather/climate related events, 39
Wood, 66–69
World Health Organization (WHO), 8
World Resources Institute (WRI), 81
2019 World Water Development Report, 71